DATA ACTION

DATA ACTION

USING DATA FOR PUBLIC GOOD

SARAH WILLIAMS

The MIT Press

Cambridge, Massachusetts

London, England

This book was set in Helvetica Neue and Minion by the MIT Press. Printed and bound in the United States of America.

Library of Congress Cataloging-in-Publication Data
Names: Williams, Sarah, author.
Title: Data action : using data for public good / Sarah Williams.
Description: Cambridge, Massachusetts : The MIT Press, [2020] | Includes
   bibliographical references and index.
Identifiers: LCCN 2019049202 | ISBN 9780262044196 (hardcover)
Subjects: LCSH: Public administration--Data processing. | Information
   behavior.
Classification: LCC JF1525.A8 W56 2020 | DDC 352.7/4--dc23
LC record available at https://lccn.loc.gov/2019049202

10   9   8   7   6   5   4   3   2   1

To my father, "the pitbull"

CONTENTS

Data
Action

INTRODUCTION: USING *DATA ACTION*

As digital systems multiply across the urban landscape, they are producing immense streams of data that can help inform how we manage and plan cities. The potential now exists, at a scale previously unavailable, to directly measure issues that have been central to urban planning since its inception, such as equity, environment, value creation, service provision, public opinion, and the effects of physical form. But many city planners find their relationship with technology, data, and its use in communities to be an uneasy one. This tension primarily results from some of the ways quantitative analysis has been misused in the past. *Data Action* seeks to respond to this problematic history by illustrating tools and techniques that demonstrate how to use the power of data to enhance learning, provoke dialogue, and inspire policy change.

Data is a medium to construct and convey ideas, just as a collection of words makes a story, or an artist who uses paint provides an image of the world. Like words on a page or paint on a canvas, a message that is shared through data represents the thoughts and ideas of the person who shares it. Data analytics and the resulting insights communicated through visualizations have done tremendous good in the world, from easing and stopping disease to exposing exploitation and human rights violations. At the same time data analytics and algorithms all too often exclude women, the poor, and ethnic groups. How do we reconcile the potential of data to marginalize people and reinforce racism with its ability to end disease and expose inhuman practices? These two realities remind us that the same data, in the hands of different people, can produce wildly different outcomes for society because how people use data shows their vision of the world. That's what makes our use of data to change the world at once exciting and alarming.

Data Action presents a corrective to standard data practices by acknowledging that data represents the ideologies of those who control its use. Data has often been used—and manipulated—to make policy decisions without much stakeholder input. *Data Action* asks advocates of big data analytics to rethink how they work with data to make the process more responsive to the people their work affects. Collaboration with policy experts, government, designers, and the public is essential to my approach: it creates trust and co-ownership in the data analysis by allowing the work to be critiqued by those who know the issue the best. The Data Action principles allow for a creative process that involves building, analyzing, communicating results, and then ground-truthing the findings through site visits and interviews—thus reworking the analytics based on the input all along the way. The results generate policy debates, influence civic decisions, and inform design to help ensure that the voices of people represented in the data are neither marginalized nor left unheard.

Recognizing that data has the potential to be used as evidence for the development of unjust policies, *Data Action* provides guidance for identifying and correcting these practices in the work we do with data. Data Action focuses on the ways we can employ data for the greater good of society, or the public good. For the purposes of this book I define the subjective concept of the public good as practices that seek to do no harm, respond to the needs of those on the margins of society, expose unjust practices, and ultimately help educate us about our world so we can make better decisions. And, although it's impossible for us to be fully unbiased when pursuing the greater good of society, the methods outlined in *Data Action* attempt to expose biases so they can be addressed directly. Much of the methods laid out in this book were developed as I pursued my goals to use data as a tool for empowerment rather than oppression. My background in geography, community planning practice, architecture, and design uniquely framed these ideas.

In *Data Action,* readers will come to recognize that using data carries responsibilities. Data is not innocent or neutral: it bears the objectives of those who handle it, and it behooves those who handle it to use it to contribute beneficially to change policy and improve their communities. Since there is no such thing as unbiased analytics, *Data Action* offers an approach that emphasizes ethical and responsible uses of data. It borrows ideas from

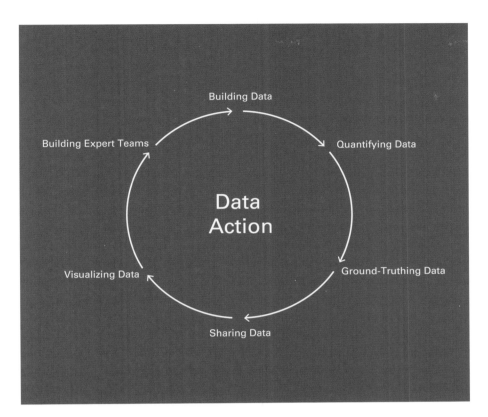

Building Data

Building Expert Teams

Quantifying Data

Data
Action

Visualizing Data

Ground-Truthing Data

Sharing Data

0.1

collaborative planning theory and practice that seek to build trust among stakeholders, help incorporate both positive and negative views, create co-ownership of the outcomes, and close the knowledge gaps among those who have power and those who don't.[1] Incorporating collaborative planning practices with data analytics sets this work apart because it includes the voices of the people in the process. This helps make the work more ethical but also builds new communities of interest through their shared work with data.

It is difficult to develop unbiased data analytics? Why? The answer starts with the existence of the data itself. Data—how it is formed and collected—cannot be separated from the underlying objectives for creating it in the first place. Bias begins there. As Lisa Gitelman argues in *"Raw Data" Is an Oxymoron*, there is no such thing as 'raw' data in the strictest sense—data bears meaning simply based on for whom and for what reason the data is created and collected.[2]

The collection and use of data has had lasting effects on our cities; the use of data in upholding segregation is perhaps the most dramatic example. Although most surveys try to eliminate biases, it is nearly impossible to eradicate them altogether. Data analysts and policy makers must be aware of how bias affects decision making in our increasingly data-driven society. Let's take, for example, the biases embedded a simple phone survey about political issues. The type of person who will answer a phone survey (i.e., the provider of the data) has certain predispositions. I, for one, am too impatient to agree to take a phone survey. The survey records the voices of those who participate, which in our example are simply people who take the time, whether they have a stake in the outcome or not. They are, in that sense, "self-selected." This can create bias. Survey designers attempt to make their questions neutral so as not to prejudice answers, but they still have agendas (and sometimes the phrasing of questions, or what issues the question addresses, betrays those agendas). Ethical surveyors try to eliminate leading questions. All too often, however, surveys are performed to support existing positions, and questions are either intentionally or unintentionally framed to generate data that support those viewpoints.

Data is used to make decisions and often reinforces the underlying objectives of those in power. Algorithms, which are increasingly used in data-generating environments, also sometimes conceal problematic biases that easily marginalize populations, for example. In her book *Weapons of Math Destruction*, Cathy O'Neil reminds us that the algorithms that run our world are developed by "fallible human beings"[3] who might have the best intentions but create results that might cause good teachers to get fired, prisoners to endure longer prison stays, or children to be removed from their families—all because an algorithm predicted these people are at high-risk of interfering with the well-being of their communities. O'Neil further makes a plea to those developing and regulating algorithms to provide ways to double-check whether the results are ethical and responsible. She asks data analysts to ground-truth their results by asking others to critique the work they have accomplished. Safiya Noble's *Algorithms of Oppression* details the dark side of data by showing how search algorithms reinforce racial and gender biases simply by yielding the results they do. In one dramatic example, searching Google for the term "black women" in 2011 returned "black women pornography" as one of the first search results. In her book, Noble explains how algorithms develop exclusionary tactics, which she calls "technological redlining," echoing twentieth-century "redlining," the systemic housing segregation of black people. In Catherine D'Ignazio and Lauren Klein's recent book, *Data Feminism*, the authors also ask readers to use data for action through the feminist lens in an attempt to break down the power dynamics.[4]

Inside Data Action

Data Action, as a book and a methodology, is a call to action that asks us to rethink our methods of using data to improve or change policy. In these pages, as the following chapter summaries explain, I provide examples from my own work and the work of others. I begin in chapter 1 by framing the discussion with a look at how we have used data since the earliest civilizations to control and manage society, with uneven results, especially for those people living on the margins.

"Big Data for Cities Is Not New"
Unjust data practices described by Safiya Noble and Cathy O'Neil in their respective works are often used by twenty-first-century cities, but this is not new. Governments of all sorts have collected data and used it as a tool of control since the earliest civilizations. According to James C. Scott, who writes about this in his 1998 book *Seeing like a State: How Certain Schemes to Improve the Human Condition Have Failed*, the collection of data by those in power was established in order to know and control the populace.[5] According to Scott, the very act of collecting data and measuring different aspects of society was established in favor of the ruling class, whether by levying taxes or establishing land rights.

In first chapter of *Data Action*, "Big Data for Cities Is Not New," I position the reader within this larger narrative by providing a historical account of the ways those in power have used data to shape our cities, how they have reinforced structural racism, and how data has helped enforce the marginalization of various populations. At the same time, I illustrate the many benefits that analyzing data has provided to society, such as establishing social services and stopping the spread of disease. The juxtaposition of these two outcomes provides a reminder that the person analyzing the data defines its use.

Historically, data has only been available to societies' elites. During the Industrial Revolution, sociologists interested in scientific methods turned to data to help address extreme poverty, lack of sanitation, and proper housing, all caused by the rapid growth of cities. Creating maps marking the location of poverty, race, cleanliness, and disease, these early planners set out to "know" their cities. These nineteenth-century sociologists began to use cadastral maps and data analysis as tools to understand the chaos created by industrialization. However, once the socio-demographic landscape was exposed, the maps were often used to create polices of exclusion and segregation, marking some neighborhoods as undesirable. The methods developed by these early sociologists still influence decision making in our cities. Today, cadastral maps are available to anyone who has a computer. Now more commonly known as parcel data, cadastral maps are openly available online in the United States, and typically can be read using open source software. While this data might be readily available for download, not everyone has the skills to use it for insights, and therefore the value of data still often lies in the hands of those in positions of power.

Ultimately, in the "Big Data for Cities Is Not New" chapter, I question our past and future motives for using data for policy change. The methodologies discussed in *Data Action* responds to such an historical account by providing guidance for working in data analytics in more responsible and ethical ways than those employed in cities in the past. Having developed the Data Action methodology over a decade of work, I illustrate it through three important calls to action: *Build it! Hack it! Share it!* In three successive chapters, I lay out the details of the Data Action method and explain my intention to inspire all of us to work with data ethically and responsibly—and for societal improvement. In each chapter, I present a data narrative from my work that embodies how the Data Action strategy works toward changing policy. Each data narrative sets the stage for other case studies that used this approach, thereby providing numerous models for those who are interested in using data for civic action.

"Build It! Data Is Never Raw, It's Collected"

It is surprising, given all the hype swirling around the massive amount of data in today's world, that some essential data is still missing. This missing data usually represents the interests and needs of those on the margins of society. But it also often represents topics that governments and companies seek to keep hidden. Chapter 2, "Build It!," challenges us—communities and data specialists alike—to create the data needed to encourage policy action. There are often dis-incentives for governments or organizations to create data themselves—they may want to keep some information purposely underexposed. "Build It!" asks us to create this data to inform the public and create civic change.

The examples in "Build It!" show that today anyone can go about collecting data with very little training: this is because innovations in digital technology make it easier than ever to collect personal data. From our mobile devices to environmental sensors in our homes, modern life is full of tools to measure aspects of our lives. It's up to us to understand the data we need, and how to use these technologies to develop narratives that can make an impact on policy. Data collection is important, and this chapter emphasizes how building data together strengthens communities around their shared interests. Building sensors, learning about data measurement, and collaborating with one another creates the bonds necessary to generate change. Whether showing the EPA that local fracking causes poor air quality and poor health or creating balloons to track oil spills, the community initiatives detailed in "Build It!" inspire us to create our own community data collection projects not only because they can literally change our community, but also because they can help create new communities.

"Hack It! Using Data Creatively"

While "Build It!" describes how to collect missing data and use it to fill gaps in knowledge, "Hack It!" (chapter 3) shows that oftentimes data already exists openly, although we might not see it that way because it is stored and maintained by private companies. The International Data Corporation (IDC) estimates that in 2025 the world will create and replicate 175ZB of data, representing more than a tenfold increase from the amount of data created a decade earlier, in 2015.[6] (A zettabyte of data is roughly equal to a trillion gigabytes, or the amount of data on 250 billion DVDs.[7]) Private companies own the majority of data existing today, and it is largely inaccessible. In some countries where data is tightly controlled by the government, privately owned data is the only data that exists openly for analyzing the dynamics of our lives. This third chapter argues that we ought to be creative in the ways we obtain data to answer important questions about society, nonetheless acknowledging that data acquired for one purpose and applied toward another holds numerous ethical concerns that must be considered.

Data Action provides guidance for those who seek to use openly available data. First we should seek to develop projects that we believe will help improve society; otherwise why take the risk of exposing people's privacy? Second, we must work with the people described in the data, not only to check the accuracy of our results but also to understand whether it might do them harm. Encouraging active participation from the groups described in the data or those who will be affected by the results allows them to co-own the data analytics. This goes a long way toward ensuring the ethical use of the data and also building action because it creates trust in the analysis and builds new communities around shared interests. Third, we must tell the people who store and control open data that we are using it, and for what purpose, to ensure it meets their terms of service; this also builds trust and shows values. Finally, we must recognize that this data is highly biased by the intention for which it was collected. For example, cell phone and internet data only represents people with those services, which means we could unintentionally marginalize the people who do not have those technologies. While most people now have cell phones, new technologies come online daily that entire populations do not have access to. So it is essential we investigate potential bias created by the technology used to collect data.

Analysis of data acquired through creative means must include subject experts and the community represented in the data. Working with policy experts helps data scientists ask the right questions. This type of multidisciplinary team can generate more accurate and ethical results from the data. There are numerous examples of data scientists who try to predict the dynamics of human life, successfully at first; but without continued inclusion of subject experts their data models are quickly outdated. This is the hubris of big data. Why do we often think the data analyst can find the right questions to ask without asking those who have in-depth knowledge of the topics we seek to understand?

"Share It! Communicating Data Insights"

"Hack It!" asks us to innovatively find, acquire, and analyze data for policy change, and emphasizes that communicating the results of those analytics ingeniously are essential for making an impact. Chapter 4, "Share It!," stresses the importance of sharing the data both in its "raw" form and also through visualizations. Sharing data helps the public have access to information, acquire knowledge, and ultimately make better civic decisions; sharing data through visualizations can communicate the insights of data without asking everyone to be a data scientist.

Sharing data through visuals is immensely effective because images allow us to quickly understand a topic, and at the same time, it is persuasive because we perceive that images hold the legitimacy of data. Human beings tend to believe data visualizations to be fact, not something to be questioned. Data visualizations also fit in our mediated culture as bite-sized, consumable thoughts that can instantly go viral on social media and the web. It not just today's current media culture that makes data visualizations so successful at communicating ideas; the process of design helps to simplify insights found in data so that anyone can understand them. It is important to remember as well that data visualizations hold the bias of whoever creates them; therefore these images reflect all kinds of political and social ideologies. I implore us all to interrogate the underlying agendas of the data visualizations we see before clicking "share" on Instagram, to prevent spreading misinformation.

Building data visualizations with collaborators, policy experts, and the community we seek to address helps to edit our biases, create trust in the message the visualization is trying to communicate, and, most importantly, disseminate that message to multiple audiences—giving the work we do with data a greater, overall impact. Such collaborative effort is the special sauce of most of the projects described in this book. In this "Share It!" chapter we look at the co-creation of data on Nairobi's semiformal transit system. This data was shared both in its raw form and through a map; the co-development created trust among the multiple stakeholders of the project—so much so that it is now the official map of the city. Sharing data in this way also allows others to acquire it and use it for their own policy change, giving the data a life beyond the purposes it was originally intended to have.

Sharing data is also important for generating political debate, which is essential to society. Data visualizations set apart a topic or an idea as something of significance or importance. As mentioned above, data visualizations work so well at communicating policy because they convey a sense of legitimacy. This is, of course, a double-edged sword as ultimately the messages data visualizations communicate come from those who design them. Being transparent, involving the public, and collaborating in the creation of data visualizations helps, but as a society we need to become more data literate so that we can critique the messages data visualizations provide.

"Data as a Public Good"

The fifth and final chapter reminds us that in a rapidly growing data landscape, there is a growing divide between the people who have access to data and those who do not. While data was once something that only landowners and governments controlled, now private companies are accumulating exponential amounts every day, which gives them the same power once held solely by governments. Some believe this amounts to *data colonialism*, where private companies extract our data as a resource and use it as a tool of control.[8] Putting the idea of data colonialism aside for a moment. I believe it is important to think of data as a *public good*: a non-rivalrous commodity that can be valued by all who consume it—a commodity similar to electricity, which needs regulations so that it can be used equitably by the public. Society should work with private companies to find ways for them to share their data ethically and responsibly so that we can use data toward the betterment of society.

Sharing data owned by private companies, however, is extremely complicated—from a property perspective and an ethical one alike. First, what incentives do companies have to share their data? After all, data supports their business, they paid for its storage and collection, and therefore it is theirs to do with as they please. When private companies do share data, they usually have an underlying agenda, which is not necessarily nefarious, but does serve their needs as much as the needs of those they share it with. Situations where the sharing of privately owned data seems the most altruistic are during a disaster events, when data, such as cell phone call detail records (CDRs), can be used to understand the spread of disease or the movement of people who need resources. Yet when data is shared this way, ethical concerns are often thrown out the window potentially putting at risk the people referenced in and affected by the data. Other altruistic data-sharing endeavors have shown that proper licensing agreements take too long to arrange, and that lag time weakens the ability to analyze data to make timely policy decisions.

While some governments are attempting to do a better job about ensuring our data privacy, such as the development of the General Data Protection Regulation (GDPR) in the European Union (see chapter 4), we still have a long way to go, as technology often moves faster than regulations. It is therefore up to those who work with data to self-regulate and ultimately try to use the data ethically. Governments, too, are asking private companies to self-regulate. But recent controversies involving Facebook and Cambridge Analytica, for instance, show there is no incentive for them to do so—and without government restrictions private companies are less inclined to monitor themselves. Having the government regulate the privacy of our data, however, seems at odds with much of the recent case law around data privacy, which is meant to protect us from government surveillance. For instance, in a June 2018 ruling for *Carpenter v. United States*, the Supreme Court declined to grant the state unrestricted access to GPS-tracked locations stored in databases of wireless telecommunication carriers.[9]

This final chapter of *Data Action* is not meant to provide answers for how privately owned data should be protected. It demonstrates instead how privately owned data, as a new resource, can do a lot of good in the world—as long as we use it ethically. I believe the Data Action methods outlined in this book help set us on the right course. Whether in the humanitarian or public health sector, or in criminal justice policy, we can collaborate in our work with data to ethically create the change we want to see in the world.

Data Action and the "Build It! Hack It! Share It!" Method

Throughout *Data Action*, I argue that unlocking data for policy change works best when the process engages multidisciplinary teams that include policy experts, data scientists, and data visualizers, among others. In the book's conclusion, "It's How We Work with Data That Really Matters" I stress that bringing together these experts allows the creative expression of data to truly blossom. Policy experts understand the issues, data scientists know how to develop algorithms, and graphic designers can share the results through compelling visuals. Working together these specialists extend their findings beyond the walls of academia or city hall—and reach the hands and minds of the public.

Data Action shows that humans are often at the center of data analytics projects, both through the algorithms they employ, the subject matter they pick to study, and the people who will be affected by the results. This means people bring their views of the world to the analytics and can create widely divergent results, both good and bad. *Data Action* attempts to create a way of handling data and data analytics more responsibly through a process of ground-truthing the results, both through human interactions and observations. It also asks us all to be critical in our approach to data analytics. Looking at the next decade, the fact that many private companies will have far more data than our governments puts these companies in a position of control. How governments worldwide work with these private institutions will be all the more crucial for ensuring that the needs of those on the margins not be forgotten.

BIG DATA FOR CITIES IS NOT NEW

Big data for cities is not new. Some might argue that data collection is as old as civilization itself. In its earliest forms, data gathered about property rights, land production, and assets was used primarily to control and extract value by those in power. James C. Scott's book, *Seeing Like a State*, reminds us that units of measurement were developed to exploit resources. For example, landowners found ways to reduce or increase the size of a bushel of wheat—perhaps by specifying whether the grain was to be heaped or leveled off, or poured from shoulder or waist height (the "longer" the pour the tighter it packed)—and they did so in their own favor as they set the standard.[1] Throughout history we have seen data used to oppress and marginalize populations, sometimes unintentionally but other times on purpose. And yet historical accounts have also shown data analytics used to improve society when applied to everything from social services to public health and eradicating disease. Ultimately, the intent of the person analyzing the data affects how it is used. Applying data to city development has presented great benefits but also posed real dangers. For that reason (at least) data analytics for city planning sparks controversy.

Data is essential for governance, of course, but in the hands of different people it will produce wildly different outcomes. This chapter provides a short history of data exploits in the city by primarily focusing on the use of data during and after industrialization, the period when society became increasingly fascinated with how scientific methods could be used to understand social processes. During industrialization we saw increases in the use of the modern census and vital statistics (population counts, births, and deaths)—all of which, being essential to governance, became a point of political debate. But the complexity of cities was often oversimplified in data models developed in postwar America, and that often led to harmful decisions. Learning from the historical successes and failures of how cities have used urban data will equip us to apply data to changing policy more ethically and responsibly. This abbreviated history of data use in cities will not only help frame how to use data for action, but also show how data can be used to wield social and political control—with sometimes devastating, other times salutary, but nonetheless lasting effects.

Data: Essential for Governance

Governing bodies of the earliest civilizations gathered data about populations in order to provide services, build infrastructure, collect taxes, and enforce policy. Historically this data was only available to the ruling class or nobility to help control the populace.[2] Ancient Rome also carried out censuses (*census* is a Latin word) as early as the fifth century BCE, and the Romans are often cited as the first to undertake a social survey. Similarly, it has been said that the earliest dynasties (and the Chinese Empire itself) would not have survived without the large bureaucracy and data collection that enabled it to control its vast territories.[3] Not surprisingly, China is home to the oldest preserved census, which in 2 CE registered 57,671,400 individuals.[4] During the Middle Ages in England, William the Conqueror (King William I) ordered all property owned by his subjects to be recorded in the *Domesday Book* (1086)—not only for the purpose of raising taxes for his army but also to determine England's wealth (figure 1.1). For the most part, data continued to appear as text in published books until William Playfair's *Commercial and Political Atlas and Statistical Breviary* (1786) introduced the "universal language" of charts and pie graphs (see figure 4.1 in chapter 4).[5] In contrast, the Incas relied on a different visual method of communication, recording census data, history, and property records in *khipu*, elaborately tied strings made of camelid hair (figure 1.2).[6]

Overall, census data today is more openly available to the public than ever before. It is used routinely by governments around the world to make decisions about everything from taxation to representation in government, to the provision of social services.[7] Still, while taking a census seems

1.1 A page from the *Domesday Book* (1086) listing land holdings in Warwickshire. Source: courtesy of J. J. N. Palmer, "Warwickshire, page 1," Open Domesday, n.d., https://opendomesday.org/book/warwickshire/01/.

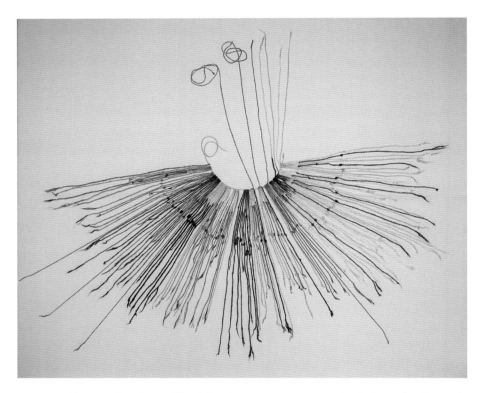

1.2 Incan fiber recording device (*khipu*) found in Peru, c. 1400–1532. Cords, knotted and twisted; cotton and wool. Overall: 85 × 108 cm (33 7/16 × 42 1/2 in.). Source: Cleveland Museum of Art, https://clevelandart.org/art/1940.469.

fundamental to governance, many countries do not conduct such surveys. Some researchers, for example, suggest that fewer than half the births in Africa in 2014 were recorded.[8] This makes planning for infrastructural improvements extremely difficult in many regions, for reasons that range from a particular government's insufficient resources to its lack of interest.[9] This is why one of the most recent United Nations Sustainable Development Goals (SDGs) includes the development of data—they want to encourage governments to collect it.[10]

The Census: Contesting Democracy through Data

Census data is a political tool. It is used to plan infrastructure, understand citizens' needs, track economic development, and more generally measure the economic and social health of communities. In *The American Census: A Social History*, Margo Anderson reminds us that the census reflects American values and the concerns of the time.[11] Every ten years, questions are added to track real and emergent trends, such as employment status, language spoken at home, or access to the internet.

Census data allows us to grasp some of the biggest unknowns in a society in order to address them through public policy. As the United States industrialized in the nineteenth century, the office of the census was professionalized. It used new statistical techniques that allowed for data sampling; and it relied on tabulating and punch card machines—invented by a census bureau worker who later founded the company that would eventually become IBM—to count responses. This was all needed to track the unprecedented growth of a country expanding in all directions. During this time, the Census Bureau also added questions about cities to reflect societal changes that were occurring as their populations doubled.[12] Census questions, whether additions or removals, tell us what governments and the public are interested in measuring, and as such reflect the values of a society.

Redistricting: The Politics of Inclusion and Exclusion

America's congressional redistricting processes, which involve redefining geopolitical boundaries, are one example of how a seemingly benign data set—the census—can produce varying outcomes, depending on the values of those who is applying the census's data. The US census was established to apportion congressional representation and determine taxes. From the very first census in 1790, the question of who should be counted caused political debate. For example, enslaved people were not recorded as full individuals but rather at three-fifths of their total. It was not until the ratification of the Fourteenth Amendment in 1868 that African Americans were counted *fully* as part of congressional redistricting.

The politics of inclusion and exclusion in the census continues today. The Trump administration proposed to include a citizenship question on the 2020 census that opponents claim will severely undercount the population, particularly in cities or areas with high numbers of immigrants who might be reluctant to be recorded.[13] According to a *New York Times* article, "if 15 percent of noncitizens went uncounted, that would be enough to cost California and New York one congressional seat each, to the benefit of Colorado and Montana."[14] The picture would be even more dramatic if only voting-age citizens were counted: California would lose five seats (figure 1.3). Government funding for services would also be affected, since federal funds are often allocated based on population figures. As of November 2018, more than two dozen states and cities had filed lawsuits against the Trump administration to get the question removed, citing that simply asking whether you are a citizen could have serious political consequences.

As of June 2019, the citizenship question was still under debate as documents emerged from the hard drive of the late Thomas B. Hofeller, a Republican operative whom the *New York Times* called the "Michelangelo of gerrymandering."[15] The documents show that Hofeller had performed a study in 2015 detailing how adding certain census questions would be to the Republicans' advantage.[16] One memo quotes Hofeller saying that adding the question regarding citizenship could be "advantageous to Republicans and non-Hispanic Whites," implying that it would reduce the number of residents who were tallied in the census from typically Democratic-voting districts with large immigrant populations.[17] This is perhaps the most current illustration of how decisions regarding data can be used to the advantage of one political ideology and party over another.

Hofeller dedicated most of his career to using data to change the political map from blue to red. In 1970, with the release of the first digital census, algorithms were used to redistrict, and Hofeller was one of the first people to apply computation models. Considered a data geek by his colleagues, he cynically supported black redistricting efforts under Section 2 of the Voting Rights Act. His goal was not to advance equal representation, something that generations of African Americans had been fighting for, but rather to clear African Americans—the most reliable Democratic voters—out of Democratic districts. Perhaps the most striking example of his intent was discovered in files found on his computer after his death that showed he used extensive profiling data to redraw congressional district lines directly through North Carolina Agricultural and Technical State University in Greensboro, North Carolina. The result divided this historically black college's campus in half, incorporating each side into majority Republican districts and thereby diluting the political voice of the A&T students. These congressional districts have since been found unconstitutional, using Hofeller's own notes as evidence.[18] Hofeller has been implicated in numerous gerrymandering cases. According to his obituary in the *New York Times*, Hofeller's creative redistricting exercises were responsible for shifting the US Congress to Republican control in 2010.[19]

Hofeller's work benefited from the lack of standardization in the redistricting process state to state, making it easy to manipulate the data toward the will of each state's redistricting committee. Officially, congressional

Congressional Seat Changes if Only Voting Age Citizens Counted

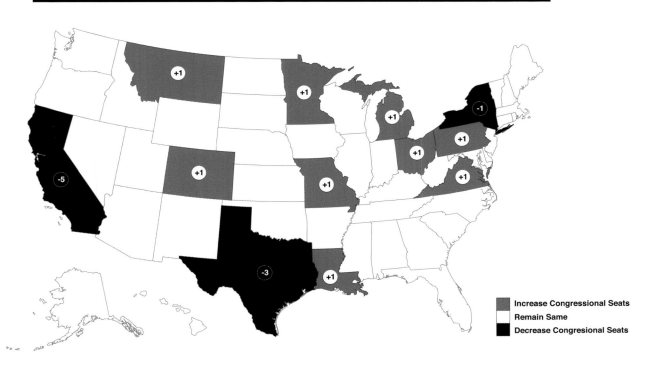

1.3 The map shows congressional seats that might be lost or won if an "only citizens of voting age" qualification were to be included in determining how congressional districts are drawn. Right now, we use the total population of each state, regardless of citizenship or eligibility to vote, to determine congressional seats. Source: Created by the Civic Data Design Lab at MIT, using data from the 2016 American Community Survey.

districts must reflect equal populations, minority representation, contiguity, and compactness, and each district must try to maintain existing political or geographic boundaries. But the decisions about how to interpret these rules rest with committees in each state whose members ultimately approve the new districts. Some states seek to preserve communities of interest (whether social, cultural, ethnic, economic, religious, or political), whereas others attempt to maintain current electoral seats.[20] But this leaves room for bias, because each state can decide how it develops its redistricting commissions. Some appoint bipartisan commissions while others select members of the majority party. States where redistricting is performed by the party in power (in Texas, for example) often see partisan gerrymandering.[21] Other states are totally inclusive (New Hampshire is one), allowing their whole legislature to make redistricting decisions. Once legislatures have redrawn congressional districts, only a court case can change any contested boundaries—and that can be a long and costly process.[22] Ultimately, redistricting shows how pushing around data can have lasting political consequences.

Fear of immigrants, represented by the emergence of the citizenship question in the 2020 census, is well documented in US political history, and according to some represents a longstanding divide between urban and rural America. In his book *Delayed Democracy*, Charles W. Eagle describes how people in the 1920s perceived rural areas as "strongholds of the Klan, prohibitionism, fundamentalism, and nativism. . . . The cities, by contrast, represented tolerance, progress, and aspects of American life." This division between urban and rural was so strong, Eagle writes, that when the 1920 census data showed for the first time how urban areas had grown larger that rural areas, the redistricting process was brought to a standstill because many rural areas feared a loss of power. It was not to their advantage to redistrict because they would surely lose congressional seats to urban areas. The standstill in redistricting meant that between 1920 and 1930, congressional districts were widely unbalanced, effectively muting this new urban population's political voice. The imbalance was eventually stabilized by the Reapportionment Act of 1929, which mandated states to redistrict in a timely manner, but questions about how to redraw district lines have remained a point of contention between urban and rural communities. For instance, in a 2017 ruling, a federal court found the redrawing of three congressional districts in Texas, including the 35th (figure 1.4), to be unconstitutional.[23] These lines were drawn to help retain congressional seats held by Republicans, who are often associated with rural America.

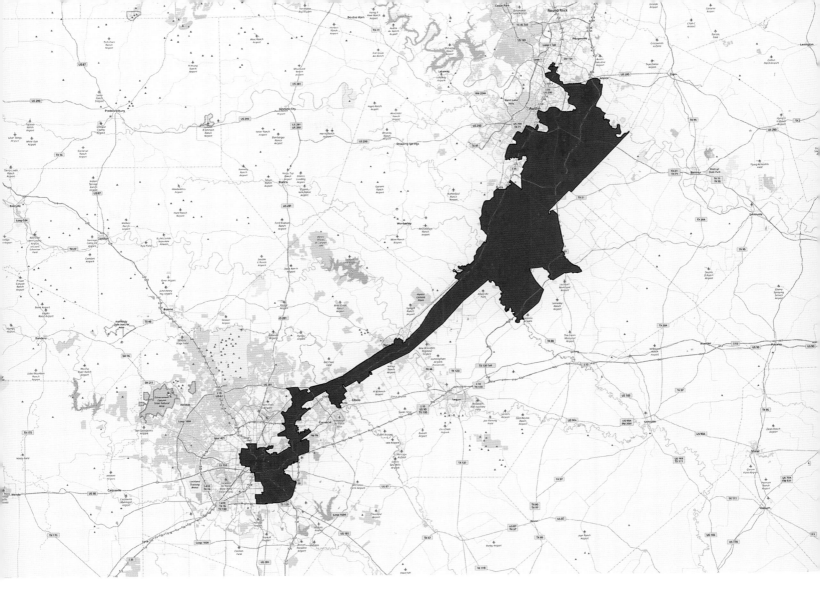

1.4 In 2017 a panel of federal judges ruled that the 35th Congressional district in Texas violated the Constitution and Voting Acts rights because race had been a motivating factor in redistricting that occurred in 2011. This district has become synonymous with gerrymandering. Source: https://www. npr.org/sections/thetwo-way/2017/03/11/519839892/federal-court-rules-three-texas-congressional-districts-illegally-drawn.

Planning for the Unknown City: Data and the Industrializing City

The rural and urban divide started when industrialization caused people to flock to cities from the countryside. This dramatic change in the character of cities created vast contrasts between the life of the rich and poor: during this time we see the emergence of vast slums and tenement housing. In many cities, these transformations put wealthy, white elites in direct contact with poverty, making them uncomfortable. At the same time, a new professional class (including medical doctors and clergymen) now had growing access to data that was once only available to government administrators.[24] Relying heavily on vital statistics available through the census, surveys, mapping, and qualitative images and interviews, some Victorian-era reformers used this new access to data in order to develop statistical techniques that would help the poor. Their work gave birth to the field of public health and the development of modern sanitation. Their outputs were often highly visual: graphs and maps helped advocate for their causes.

Data Analytics: Giving Birth to the Field of Public Health

The First Industrial Revolution (1760–1840) transformed the dependency on manual labor in the economies of societies to the deployment and reliance on steam-powered machines. Urban populations grew tremendously as people moved from the countryside seeking work in factories. London, perhaps the first mega-city, doubled its population between 1802 and 1850. With that growth came extreme poverty, a lack of proper housing and sanitation, and outbreaks of cholera (in 1832 and 1850s).[25] Such outbreaks, it turns out, propelled early nineteenth-century data collection efforts, often in direct response to public health issues, living conditions, or poverty.[26] In Britain, this interest resulted in the establishment of a number of statistical societies,[27] the first of which, the Manchester Historical Society, was founded in 1833, at approximately the same time that the United Kingdom's Ordnance Survey started to create cadastral (detailed map) surveys of cities.[28] Just a year later, in 1834, Charles Babbage co-founded the Statistical Society of London, which later became the Royal Statistical Society. Increased access to data helped encourage the development of these new organizations to address issues important to society.

The fields of epidemiology and public health were born from this new, heightened interest in statistical analysis and mapping. Edwin Chadwick, John Snow, and William Farr, among other notable public health statisticians, were members of these societies. Their sharing of data and their analytical methods generated groundbreaking research and helped eradicate disease. Most notably, perhaps, they shared their data through visualizations that communicated their ideas clearly enough for the results to be effective at

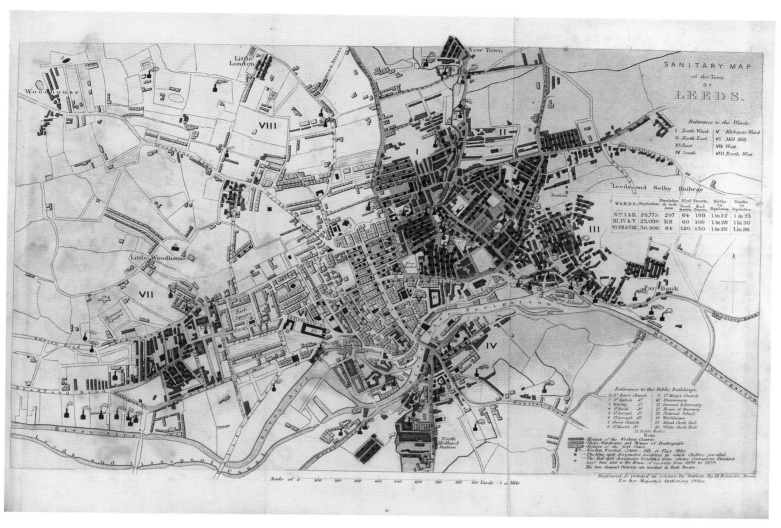

1.5 Sanitary map of the Town of Leeds from the report of Sanitary Condition of the Labouring Population of Great Britain (1842). Source: "1842 Sanitary Report," https://www.parliament.uk/about/living-heritage/transformingsociety/livinglearning/coll-9-health1/health-02/1842-sanitary-report-leeds/.

changing policy. Chadwick developed the *Report on the Sanitary Condition of the Labouring Population* (1842), which graphically described the physical and social concerns that might be contributing to outbreaks of cholera (figure 1.5). The compelling findings and the ease with which the data was communicated through maps generated debate that is said to have helped lead to the passage of Britain's Public Health Act in 1848.[29]

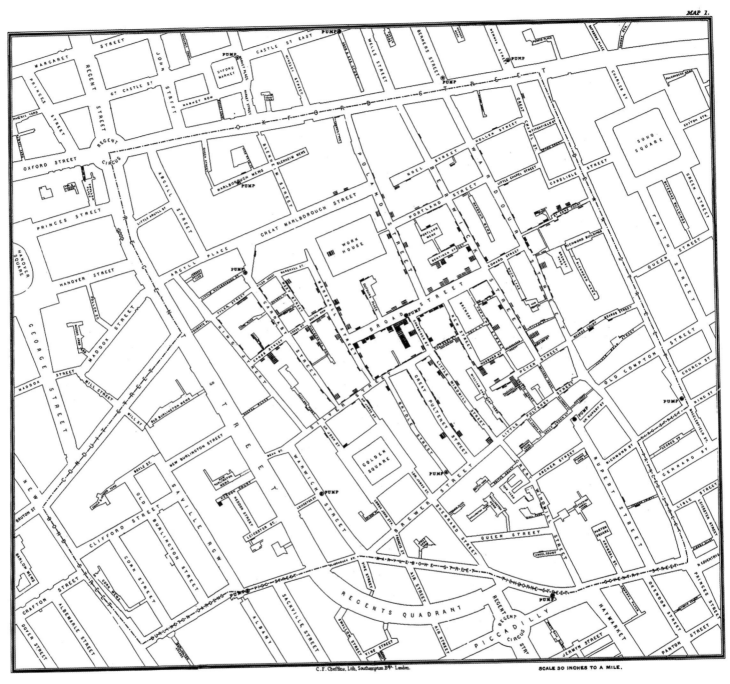

MAP 1.

SCALE 30 INCHES TO A MILE.

C.F. Cheffins, Lith. Southampton B⁹. London.

1.6 John Snow's cholera map (1854) shows the high numbers of deaths from cholera around the Broad Street water pump. The map is often considered an example of data visualization because it included bar charts to pinpoint the number of deaths at a particular address. Source: John Snow, *On the Mode of Communication of Cholera* (London: John Churchill, 1855).

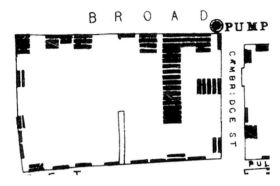

1.7 Zoom of the bar charts used on John Snow's cholera map (1854), shown in figure 1.6.

John Snow developed his now famous map (figures 1.6 and 1.7) of the Broad Street Pump (1855), which linked deaths from cholera to the contaminated pump water. This could only be done after William Farr, Compiler of Abstracts (or chief statistician) for Britain's General Register Office (GRO), openly shared with Snow statistical data on deaths from cholera. Snow's maps showed the geographic concentration of cholera deaths to those households who used the Broad Street pump as a water source.[30] Before Snow's mapping experiment, scientists did not widely accept that cholera could be carried in water, but the maps provided overwhelming visual evidence of the need to revise this assumption. Snow's visualizations have had a lasting influence on the fields of public health and sociology because of their ability to communicate information essential to policy needs and policy-making. Farr, who was the equivalent of a present day chief technology officer, went on to develop the statistical evidence for Snow's maps, showing that his work was neither an anomaly nor visual artifice.

Farr's tenure (1851–1871) overseeing Britain's censuses and his role as Britain's chief statistician might be considered the first open data gesture: it marks an important, early moment in the growth of Britain's open data movement. Under his direction, the types of data available in England multiplied and began to include such vital statistics as place of birth and registration of deaths (1841).[31] Farr performed extensive studies using death certificates, looking at the rates of suicides among different classes and education levels, occupational mortality, and causes of death. As he became interested in the spread of disease he studied influenza (1847) and cholera (1854) in particular.[32]

Cadastral Mapping: Controlling the Industrialized City

Data in the form of maps became a communication tool for Victorian-era social statisticians to document land assets, poverty, and "subversive elements" of the city. By highlighting details such as the locations of brothels or of immigrants with distasteful social norms, these maps became a tool for influencing policies to control unsavory elements of the city. Although many of these maps were conceptualized to address the plight of the poor, once put to use they often further marginalized them, simply by marking them as "other" on the map. These early maps illustrate the duality that exists within the application of data for policy change. Data can be used as evidence that can help or harm.

Victorian-era fascination with maps can be seen as early as 1824, when Britain's Ordnance Survey began developing detailed maps of cities and towns, showing for the first time building footprints, streets, roads, parks, and civic institutions such as churches, schools, and government buildings. These maps were largely used for taxation purposes. In 1862, the Ordnance Survey released one of the largest and most detailed in this series: London at a scale of 1:10,560 (where 6 inches on the map represents a real-world distance of 1 mile).[33] In the United States, private companies made similar maps; for instance, in 1867, the Sanborn National Insurance Diagram Company (later called Sanborn Fire Insurance) released its first map of Boston, which included incredible detailing of brick and wooden structures, contaminants, and types of businesses. The company went on to develop the same maps for most American cities. These early maps are still used today for environmental remediation because they include important data about where oil or other chemicals were used, thereby helping to construct an environmental history of cities (figures 1.8 and 1.9). Employing an army of cartographers for more than one hundred years, the Sanborn Company became the only source of data that traced the vast and rapid development of US cities.

Sanborn maps and their London cousins, the Goad Fire Insurance Plans (1886–1930), became an invaluable resource for land use information. Because they were focused on the built environment, however, the maps did not measure socioeconomic conditions of the city. Charles Booth, a London philanthropist, set out to change that. He believed a survey of poverty on the scale of these maps would be essential to study the condition of the poor.[34] This led Booth to conduct the most extensive survey on race and class London

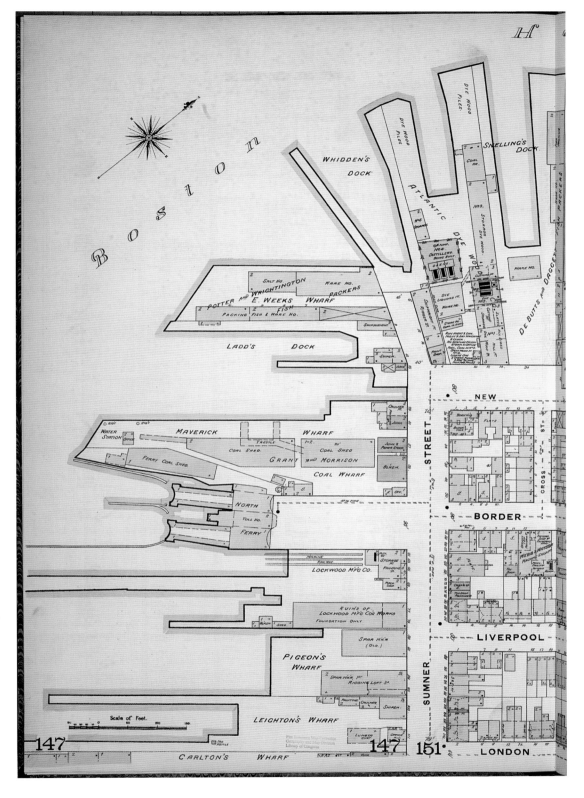

1.8 This Sanborn map from 1888 shows a block in a manufacturing district in Boston, Massachusetts, including a coal shed, dye factory, and machine shops. The maps are an example of some of the first detailed data sets on cities. Source: Sanborn Map Company, "Image 3 of Sanborn Fire Insurance Map from Boston, Suffolk County, Massachusetts," 5 (1888), http://hdl.loc.gov/loc.gmd/g3764bm.g03693188805.

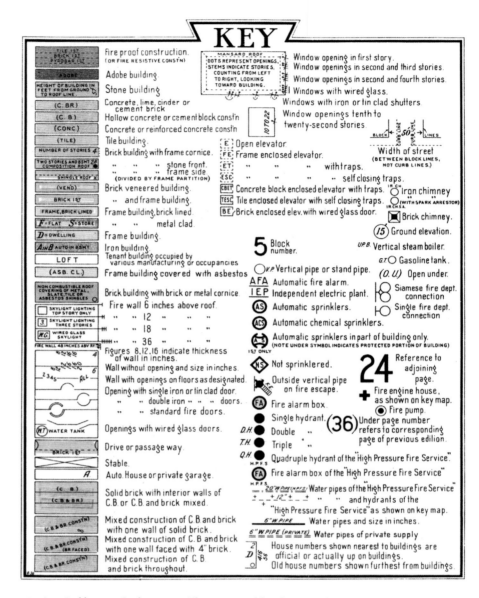

1.9 A typical key to a Sanborn map. The amount of detail is more than cities collect about most properties today. Source: FIMo—How to Interpret Sanborn Maps, 2019, http://www.historicalinfo.com/fimo-interpret-sanborn-maps/#Utilization.

had ever seen. From 1892 to 1903, Booth's team went to every property in London and asked a series of questions about income, quality of housing, and access to water, among other conditions of the urban environment.

Booth's survey was translated into the now-famous *Descriptive Map of London Poverty* (figures 1.10a and 1.10b), which comprised the first ever sociodemographic maps of London.[35] Booth's map helped to properly estimate the poverty level, claiming that 30.7 percent of London's population lived in poverty.[36] Previous estimates were closer to 25 percent, so this represented a marked difference.[37] *The Maps* were considered exceptional insofar as they were the first to link social issues with the physical city; Booth's work helped philanthropists and governments to understand many social–spatial aspects of the lives of the urban poor in new ways (figure 1.10).[38] Although social surveyors of the time often employed tables, numbers, and statistics, Booth's maps were cited as proffering objective truth, adding credibility to his findings.[39]

At the same time, Booth's work became a tool to advance the systemized removal of whole populations from London's inner core; for example it inspired the development in 1901 of the Royal Housing Commission, where the maps were used to rationalize the complete demolition of slums and the rebuilding of social housing in their place. Nichol housing (a slum in Jewish East London), for example, was replaced by Boundary Estates, the first state social housing in Britain.[40] A housing upgrade may indeed have been needed in East London, but those who lived in the Nichol slum could not afford the rents for the new estates, which were set higher than what they had paid for Nichols.[41] As a result, an entire Jewish community was pushed out of their neighborhood. Looking back at the records, it is difficult to determine whether this was a purposeful targeting of a particular group. According to some accounts, Booth was "confused" by the Jewish culture and in some of his writings appears to be anti-Semitic.[42] This possible prejudice may have led to his focus on the Nichol area, although it appeared to have been slated for redevelopment before Booth's maps were available. Therefore, the maps may have simply created rational evidence to garner public support for policies the city government already wanted to enact.

Philanthropists in America and Europe imitated Booth's work, and although many of the maps generated evidence for projects that helped the urban poor, others often had racist undertones and were used as evidence to enact discriminatory policies.[43] For example, in1885 San Francisco's Board of

1.10a and 1.10b Charles Booth's Social Demographic Maps of London. Figure 1.10a shows the original map sheet 5, focusing on East London, which was home to many Jewish communities at the turn of the twentieth century. Figure 1.10b is an enlarged view of the same map focused on the former location of St. Nichol slum, which was cleared to create Boundary Estate, one of London's first social housing experiments. Source: London School of Economics and Political Science, "Charles Booth's London," accessed August 2, 2019, https://booth.lse.ac.uk/map/13/-0.1565/51.5087/100/0.

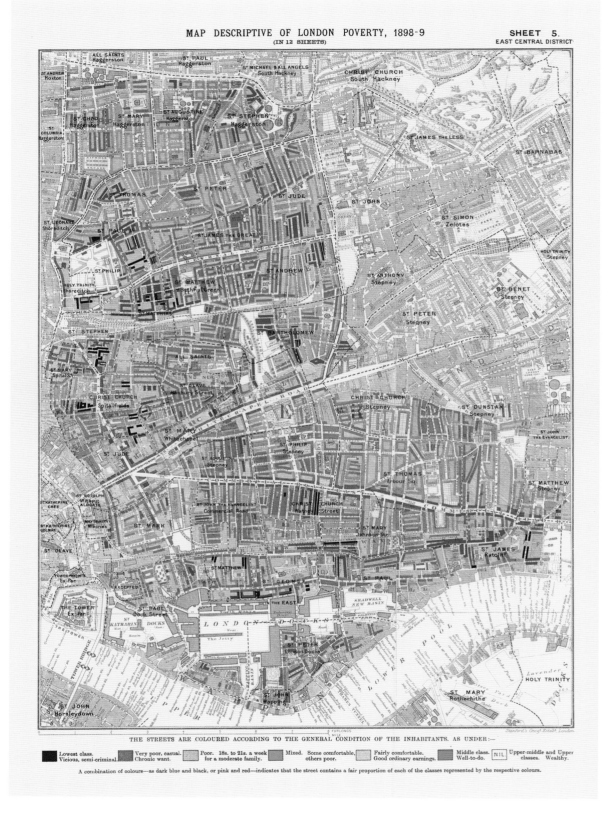

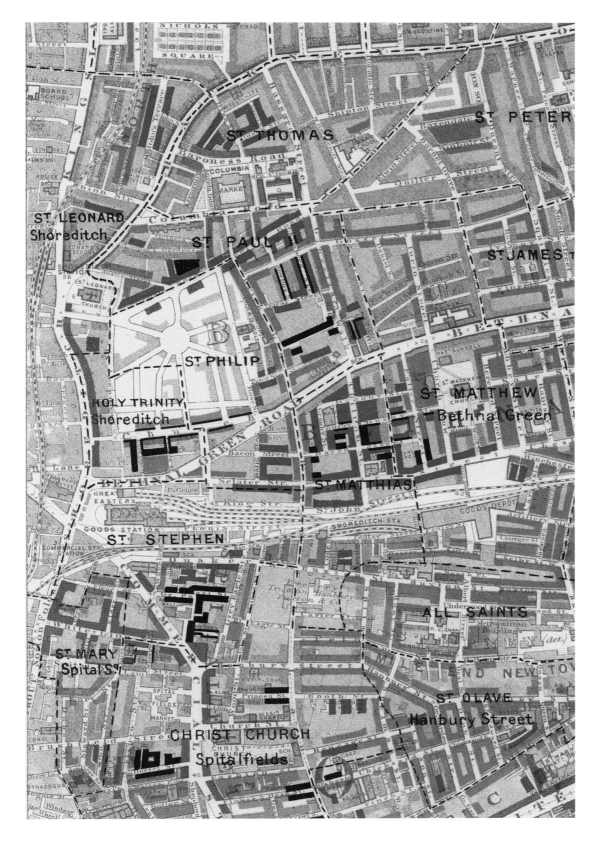

Supervisors released a map of Chinatown's gambling, drug, and prostitution businesses (figure 1.11). The report that accompanied the map, which was written in the course of developing policies that would remove Chinese residents from the area, states: "We have permitted the Chinese to become our masters, instead of asserting and maintaining the mastery ourselves."[44] As told in Nayan Shah's book *Contagious Divides*, "nineteenth-century San Francisco health officers and politicians conceived of Chinatown as the preeminent site

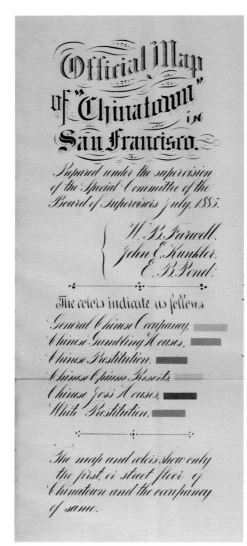

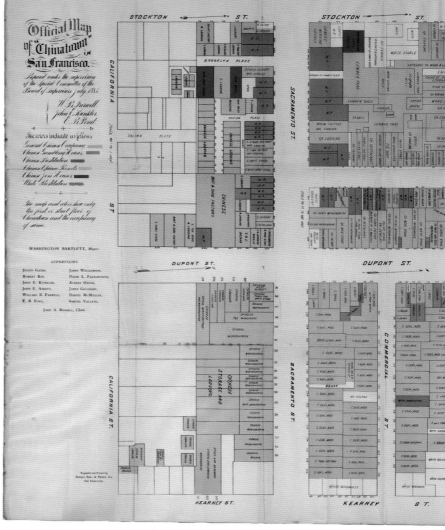

of urban sickness, vice, crime, poverty, and depravity."[45] Shah interprets the maps (figure 1.11) as surveillance, and their continued existence ensured this type of surveillance in the future. Maps became important tools for enacting city policies and were oftentimes used to segregate populations seen as "other." Where some employed cadastral maps to marginalize populations, many used them to expose and improve the quality of life of the urban poor. These types of cadastral maps were also used by the "settlement movement"

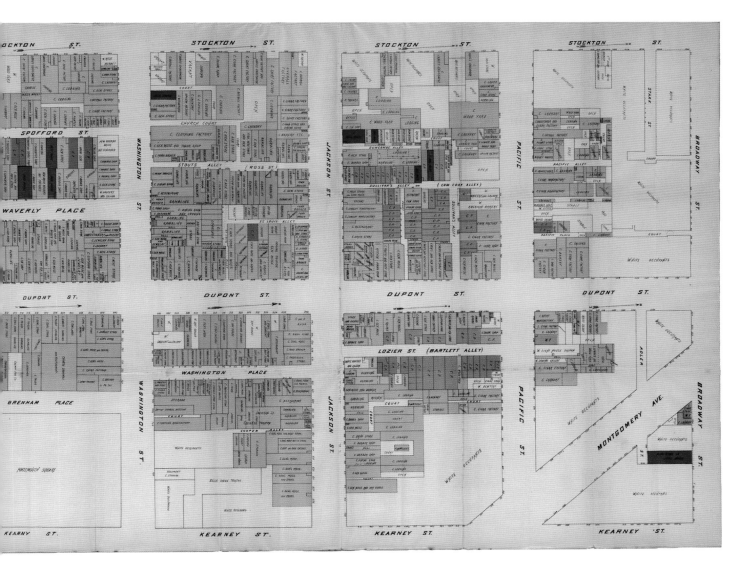

to help the urban poor. Chicago's Hull House is perhaps the most famous of these settlement houses, which provided services such as childcare, education, and healthcare. In 1895, Jane Addams, who ran Hull House, and her partner Florence Kelley used Booth's London project as a model for the *Hull House Maps and Papers*. The maps helped to show that rents were not at all proportional to wages—helping to expose how landlords preyed on the urban poor (figure 1.12).[46]

Addams and Kelley's work in turn influenced W. E. B. Du Bois's well-known study of the African American population in Philadelphia, released as *The Philadelphia Negro* (figure 1.13).[47] Du Bois extensively surveyed more than 5,000 residents regarding wages, employment, education, and housing stock, among other things. He noted that many African Americans lived in conditions that made their rise out of poverty difficult (figure 1.14).[48] Du Bois used the strategy in his other works, including his studies for the Conferences of the Atlanta Negro (figure 1.15).[49] In both Addams's and Du Bois's work, data visualizations in the form of maps helped create evidence for the existence of structural issues surrounding poverty, allowing viewers to reflect on those issues. How could an individual be expected to rise out of poverty when their wages were lower than even the most basic living expenses? Early reformers helped to raise awareness of this problem among broader groups of people.

It should be noted that at the same time that Jane Addams and W. E. B. Du Bois were working in Chicago, academics at the University of Chicago were also compiling qualitative data on the urban poor, eventually forming what became known as the "Chicago School" of sociology (1915–1935). Led by Robert E. Park and Ernest Burgess, the Chicago School saw the city as its urban laboratory and used the same methods of participatory observation that anthropologists were beginning to use with remote, indigenous populations.[50] The Chicago School's methods—which included data analytics, mapping, and qualitative research (figures 1.16 and 1.17)—were not meant to solve the problems of the city, but to try to define and describe the modern city, empirically, creating a "science of society." Their extensive mapping exercises led Burgess to create his famous "concentric zone" diagram (figure 1.16), which he used to explain racial segregation as a natural progression of city development, and therefore something that was unchangeable. These theories stood in stark contrast to how Jane Addams and W. E. B. DuBois used the same mapping techniques as tools to change economic and labor policy. The maps of the Chicago School are controversial because scholars believe

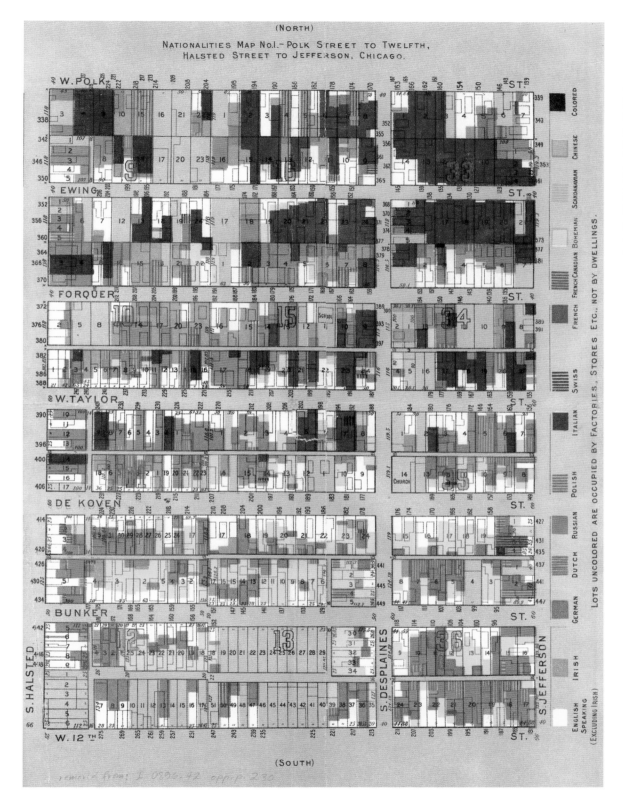

1.12 Map from Hull House Maps and Papers (1895) shows the different nationalities in an immigrant neighborhood around Hull House. Source: "Maps of Gilded Age San Francisco, Chicago, and New York," *Mapping the Nation* (blog), accessed June 29, 2019, http://www.mappingthenation.com/blog/maps-of-gilded-age-san-francisco-chicago-and-new-york/.

1.13 W. E. B. Du Bois included this map in The Philadelphia Negro (1899), an extensive survey of African American in the city's seventh ward of Philadelphia. The green and red in this map shows "working class" and "middle class." Source: William Edward Burghardt Du Bois and Isabel Eaton, The Philadelphia Negro: A Social Study, 14 (published for the University, 1899).

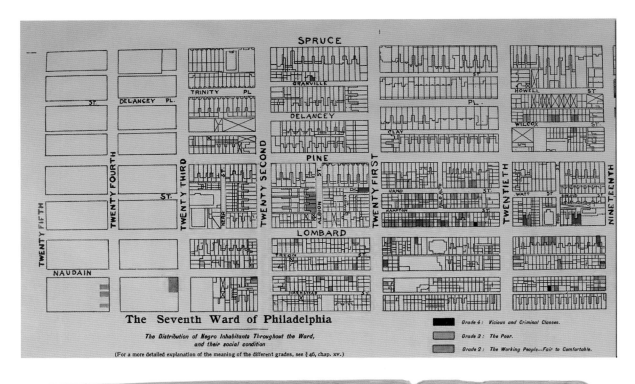

1.14 W. E. B. Du Bois's chart shows the average household budget of African Americans in Atlanta. Source: W. E. B. Du Bois, [The Georgia Negro] Income and Expenditure of 150 Negro Families in Atlanta, Ga., U.S.A., circa 1900, drawing: ink, watercolor, and photographic print, 710 x 560 mm (board), circa 1900, LOT 11931, no. 31 [P&P], Library of Congress Prints and Photographs Division, https://www.loc.gov/pictures/item/2013650354/.

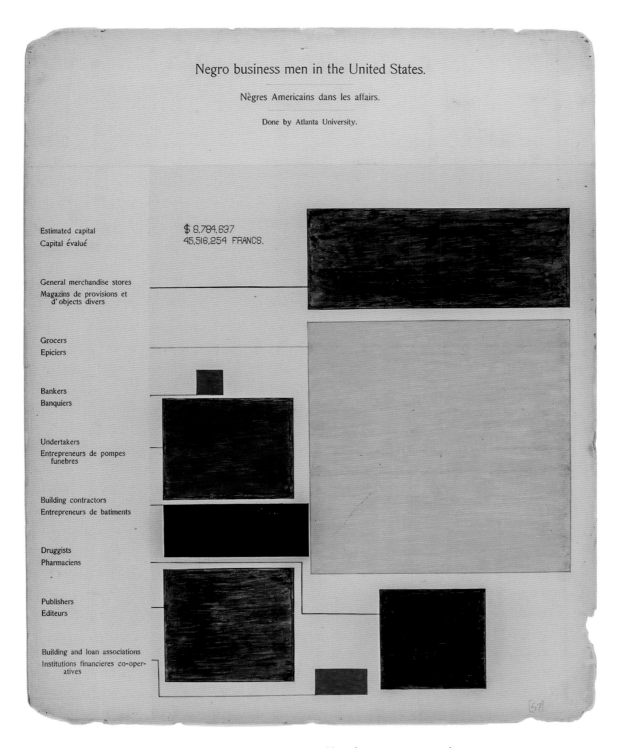

1.15 W. E. B. Du Bois's chart shows the types of businesses owned by African Americans in the United States, illustrating that black men had jobs similar to those of white men and helping to elevate the status of African Americans. Source: W. E. B. Du Bois, [A Series of Statistical Charts Illustrating the Condition of the Descendants of Former African Slaves Now in Residence in the United States of America] Negro Business Men in the United States, circa 1900, 1 drawing: ink and watercolor, 710 x 560 mm, circa 1900, LOT 11931, no. 57 (M) [P&P], Library of Congress Prints and Photographs Division, https://www.loc.gov/pictures/item/2014645363/.

1.16 The first hand drawn map of Burgess's famous theoretical model used to explain the social organization within urban areas. The diagram is central to the Chicago School of Sociology's work. Source: Ernest Watson Burgess, "Map of the Radial Expansion and the Five Urban Zones," n.d., Ernest Watson Burgess Papers, University of Chicago Library.

1.17 This map accompanies Frederic Thrasher's *The Gang: A Study of 1,313 Gangs in Chicago*, a text about the Chicago School of Sociology. This map was used to argue that it is not just geography that causes gang behavior but also the breakdown of institutions. Source: Frederic M. Thrasher, *The Gang: A Study of 1,313 Gangs in Chicago*. Chicago: University of Chicago Press, 1927.

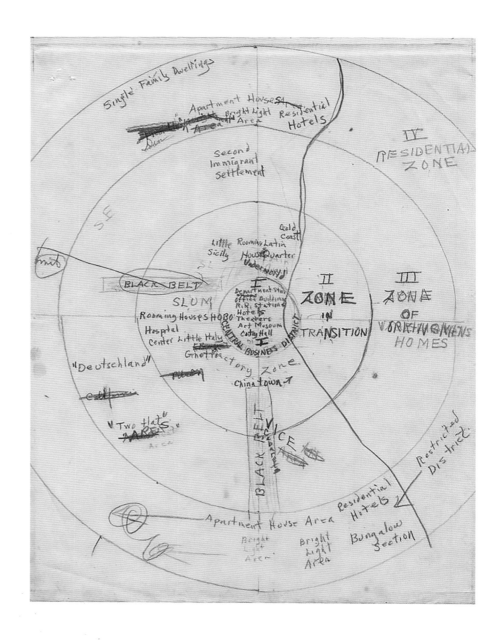

CHICAGO'S
GANGLAND

PREPARED BY
FREDERIC M. THRASHER
1923–26

LEGEND
▲ Gangs with Clubrooms
● Gangs without Clubrooms
Parks, Boulevards, and Cemeteries
Industrial Property
Railroad Property

1.18 The authors of the Pittsburgh Survey developed multiple graphic devices to share their work. Source: Paul Underwood Kellogg, "Figure 1.1," in The Pittsburgh Survey: Findings in Six Volumes, ed. 1909 (112 East 64th Street, New York, NY 10065: © Russell Sage Foundation, n.d.).

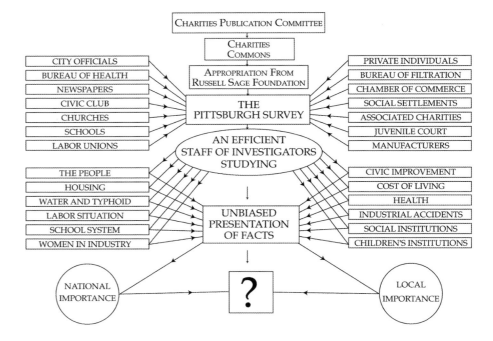

1.19 Maps and photographs were an important part of the Pittsburgh Survey. Source: Paul Underwood Kellogg, The Pittsburgh Survey; Findings in Six Volumes, vol. 5, 1909, https://archive.org/details/pittsburghsurvey05kelluoft/page/xvi. Source: Paul Underwood Kellogg, "Figure 1.1," in The Pittsburgh Survey: Findings in Six Volumes, ed. 1909 (112 East 64th Street, New York, NY 10065: © Russell Sage Foundation, n.d.).

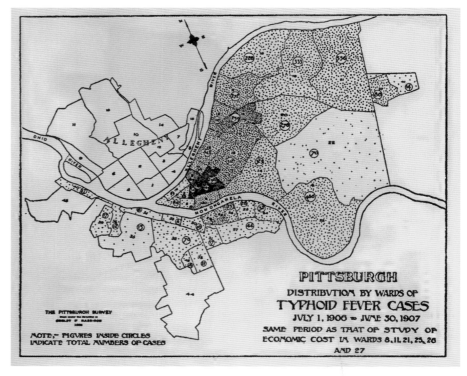

1.20a and 1.20b Photographs taken during the Pittsburgh Survey. Source: Margaret Frances Byington, Paul Underwood Kellogg, and Russell Sage Foundation, *Homestead: The Households of a Mill Town*, 1910, https://archive.org/details/homesteadhouseho00byinuoft/page/n189; Crystal Eastman, *Work-Accidents and the Law*, 1910, https://archive.org/details/cu31924019223035/page/n231.

they reinforce ideologies steeped in structural racism. What is perhaps more problematic is that even though the Chicago School learned its mapping techniques from Addams and DuBois, it marginalizes their connection to this earlier work by a woman and an African American sociologist.[51]

The social work being done to understand the poor in Chicago (by both Addams and the Chicago School of Sociology) influenced fields beyond sociology and was widely adopted by city planners. This work also influenced philanthropic organizations. The Russell Sage Foundation in particular funded extensive surveys on the social conditions in numerous US cities. Among the most notable was its *Pittsburgh Survey* in 1907 (figures 1.18 through 1.20b), which collected extensive data on living conditions, sanitation, light and air, education, and poverty, among other characteristics.[52] Hiring specialists from numerous fields, Russell Sage researchers made use of photographs, maps, and charts to communicate their results, and set the standard for others gathering qualitative and quantitative data to tell the story of the industrialized city. Data, both qualitative and quantitative, became crucial to changing the quality of life of people in cities everywhere.

Cadastral Mapping: Modern City Planners

At the turn of the twentieth century city planning had yet to become a profession, so city planners were often trained as civil engineers. They took cues from the surveys performed by sociologists such as Addams, Du Bois, the Chicago School, and the Russell Sage Foundation to develop zoning and land-use plans that they hoped would harness the benefits of rapid urbanization. New York City's first zoning plan was established in 1916 and set out to separate incongruent uses. The new plan zoned Midtown West for manufacturing (largely of apparel) to provide a location separate from tenement housing. Early engineers, who were often referred to as technocratic planners, used detailed surveys to map out their cause, just as Addams and her colleagues did. Their objectives, however, often differed because they were largely hired to help increase the economic prowess of cities rather than address their social needs, such as improving poverty conditions. In his book *The Birth of City Planning in the United States*, Jon Peterson explains how this marks a shift from planning ideals that come from moral reform to those based on scientific progressivism.[53] Technocratic planning is based on a conception of decision making anchored in objective knowledge, such as

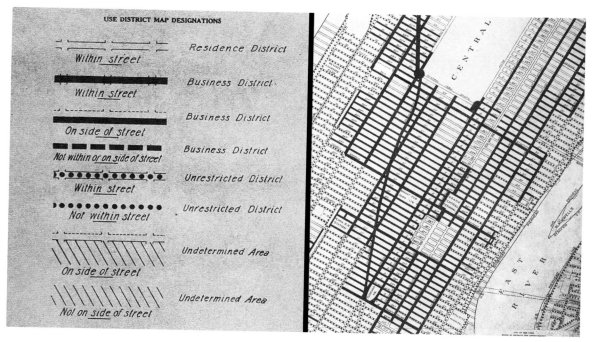

1.21 City of New York Board of Estimate and Apportionment, Use District Map, 1916. This is considered one of the first zoning maps. Source: https://digitalcollections.nypl.org/items/510d47e4-80df-a3d9-e040-e00a18064a99.

data, that will yield maximum benefits for the public.[54] It involves collecting extensive data and using it as evidence to develop zoning regulations to shape the city (see an example of zoning use designations in New York City in figure 1.21, considered one of the first zoning maps).

Harland Bartholomew professionalized this new scientific approach that introduced zoning as a main tool for land-use planning. He relied on data to develop his planning ideas, using it as evidence to support his own beliefs and ideologies—which, to put it mildly, often lacked concern for those on society's margins. For Bartholomew, the economic health of the city was the most vital concern. He believed removing the blighted areas of the city and improving access to its central business district would help the city's economy and increase its overall wealth, which in turn would be good for everyone.[55] This vision is reflected in his numerous plans. They emphasize slum removal, the development of highways to enhance efficiency over maintenance of

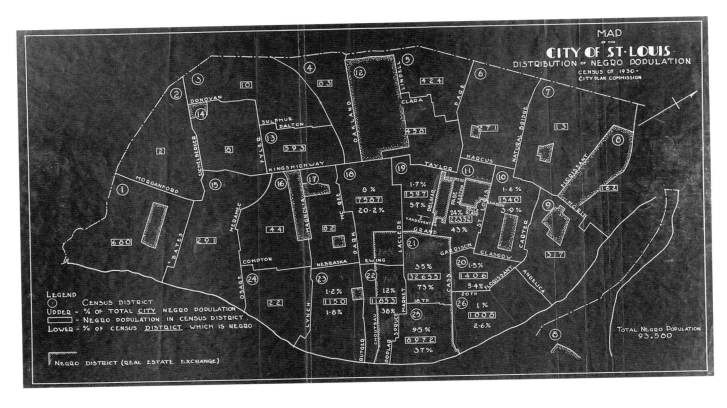

1.22 This 1934 map, based on the 1930 census, shows both the distribution of the black population and the boundaries of the St. Louis Real Estate Exchange's 1923 "unrestricted" zones. Source: City Plan Commission, "Map of the City of St. Louis: Distribution of Negro Population, Census of 1930 (Realtors' Red Line Map) 1934, http://collections.mohistory.org/resource/221591.

existing communities, and a focus on development in business districts in the core of the city. His analyses often specifically mention the need to remove or marginalize African American neighborhoods.

Bartholomew's approach, especially in the city of St. Louis, Missouri, illustrates the problematic ways in which city planners have employed data in the past, a legacy that helped facilitate the systematic segregation of US cities. Bartholomew became St. Louis's first city planner in 1915 and served in that capacity until 1950. He began his work there by implementing extensive citywide surveys of the character and use of buildings, using the Sanborn maps as his base. His surveyors visited each property and collected detailed information on its use and ownership—including the race of the owner(s) and/or occupant(s) of each residential building in the city.

Bartholomew used this data on race primarily to determine which neighborhoods were largely populated by African Americans, and then zoned those neighborhoods for undesirable uses, such as brothels, liquor stores, and manufacturing, helping to encourage blight in once middle- and lower-class African American neighborhoods. Many found that the zoning caused their property values to significantly decrease; as a result, when African Americans wanted to move, they could not afford property in equivalent middle- or lower-class white neighborhoods (figure 1.22). In his own words, his goals were to preserve "the more desirable residential neighborhoods" and to prevent the movement into "finer residential districts . . . by colored people."[56]

Bartholomew used the data to develop the zoning ordinance of 1919, which slated areas in and adjacent to primarily "colored" communities for industry and manufacturing, making them less attractive to middle-class residents who were predominantly white.[57] The ordinance doesn't specifically mention race because zoning based on race was indeed illegal at this time, but the results of the new ordinance created a segregated city. African American communities were pushed to the margins. What is more problematic, using this plan as an example, the St. Louis zoning commission often rezoned neighborhoods for industrial use if it saw African Americans moving to those locations (figure 1.23).[58] This practice would make it hard for potential residents to receive mortgages, helping to further blight these communities.

1.23 Bartholomew classified many African American neighborhoods in St. Louis as industrial districts. This photo shows a factory located right next to housing. Similar juxtapositions caused many of these neighborhoods to face severe decline. Source: Harland Bartholomew and City Plan Commission, Zoning for St. Louis: A Fundamental Part of the City Plan (St. Louis, Mo., 1918), https://ia902708.us.archive.org/5/items/ZoningForSTL/Zoning%20for%20STL.pdf.

The St. Louis zoning commission continued to use these exclusionary practices under Bartholomew's lengthy tenure as chief planner and beyond. By allowing land use that was not permitted in other residential neighborhoods—polluting industries, taverns, liquor stores, nightclubs, and brothels—the segregation helped turn these African American neighborhoods into slums.[59] African American homes in these neighborhoods were not eligible for Federal Housing Commission mortgages because their terms did not provide for incongruent use. Ironically, these zoning and use designations helped St. Louis garner urban renewal funds that paid for the ultimate clearing of many of these communities, which in one case led to the design and construction of the infamously dysfunctional Pruitt-Igoe housing project.[60] Some might argue that the clearing of these communities was always the goal of St. Louis's exclusionary zoning practices.

During the course of Bartholomew's career his firm developed over 563 comprehensive plans and his books and lectures influenced both the professionalization and the evolving profession of planning.[61] His methods, detailed in *Urban Land Uses* (1932, updated 1955), reached a broad audience, and the book became the standard text for academic planning programs across the country.[62] Using data resources amassed by Bartholomew's firm during the time it took to develop these numerous plans, the book provided a comparison of land-use and zoning practices (and their outcomes) in cities across the United States. Bartholomew also lectured at influential planning schools including Harvard and the University of Illinois, which helped to further circulate his ideas. The technocratic field of planning that Bartholomew helped to develop later became synonymous with the marginalization of the urban poor, giving rise to distinct practices of racial segregation.

Cadastral Mapping: Groundwork for Lasting Oppression

The types of cadastral maps and zoning ordinances developed by Bartholomew's firm laid the groundwork, literally and figuratively, for the practice of "redlining," or discriminatory lending practices based on the geographic location of properties. Community groups in Chicago's Austin neighborhood in the 1960s coined the term "redlining" in reference to the red lines that lenders and insurance providers admitted to drawing around the areas where they would not provide mortgages.[63] Bartholomew's zoning maps provided the impetus for disinvestment in African American communities that the Home Owners' Loan Corporation (HOLC) maps (1933–1936) continued (figure 1.24). HOLC maps identified mortgage risk in 239 cities

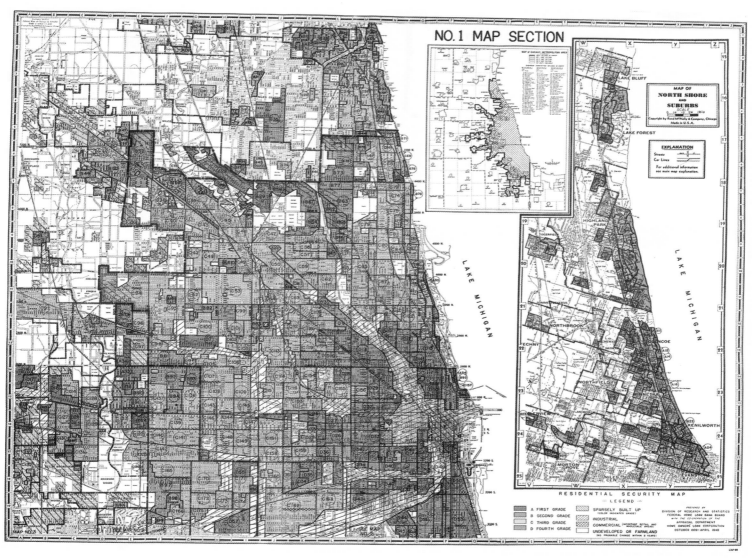

1.24 Home Owners' Loan Corporation maps of Chicago, showing the inner-city African American neighborhoods in red. Source: Form-Based Codes Institute, "Zoning for Equity: Raising All Boats," 2019, https://formbasedcodes.org/blog/zoning-equity-raising-boats/.

across the United States, using four categories. The highest risk—the red color on HOLC maps—identified African American communities, areas with so-called incongruent uses and poor housing stock.

The base data that HOLC used was developed from the 1934 and 1939 Works Progress Administration's Real Property Surveys, which mapped major cities across the United States, and which provided jobs during the Great Depression. The process of making the HOLC maps was overseen by the HOLC's Department of Economics and Statistics, which asked residents questions directed to identify risk, such as whether they held mortgages or felt they could hold one. Since Bartholomew was involved with the development of this program, many of the detailed, data-collecting strategies he used for his own work were used here too. According to Amy E. Hillier, these maps were given to mortgage lenders to identify areas where loans should not be given. The maps had a huge impact on helping to further marginalize these neighborhoods.[64]

Redlining maps disguise visible structural bias as objective truth. They also illustrate how the decision-making practices of banks were typically racially biased. These maps have become important symbols for how visualizations can be used powerfully, often by the powerful, to oppress the powerless. These maps permanently marked these urban landscapes as "other," places not worth investing in, and the ramifications of that labeling can still be seen today.

Data and Urban Renewal

Technocratic planners continued to apply data and mapping to urban renewal policies in the 1950s, securing funds from federal highway and housing grants (the 1949 Housing Act, for example) that could be gotten only if data were submitted as evidence in the application for funding. In cities across the country, technocratic planners devised ways to send highways through thriving neighborhoods they designated as "blighted" in the hopes of permanently removing these areas, using data as evidence. Perhaps one of the most notable occurrences of this phenomenon—the changes made to the original Interstate 95 highway plan for Miami—resulted in the total obliteration of Overtown, the city's historic African American neighborhood.

In 1955, Miami transportation planners set the course for the I-95 highway route to be run along an abandoned rail corridor, on the edge of a residential area, and through "low-value" industrial tracts "to preserve and help protect existing residential neighborhoods and promote an economically desirable use of land."[65] But a year later, the highway route was changed when the state of Florida hired outside consultants—Wilbur Smith and Associates, a firm that had designed many other interstate routes across the country. The firm's engineers moved the route several blocks west to run through the inner core of Overtown, often called the Harlem of the South because of its lively music scene; they rationalized that the shift provided "ample room for future expansion of the [city's] central business district in a westerly direction."[66] The new plan had irreversible impacts on the city—the total removal of an African American cultural district.

Miami's business community did not see Overtown as a community that contributed to the economic viability of Miami (figure 1.25); they instead saw

1.25 This street scene of Overtown shows African American shoppers outside the retail establishments along NW 2nd Avenue between 8th and 9th Streets, circa 1950. Miami News Collection, HistoryMiami Museum, 1989-011-1703.

1.26 Massive highway interchange clears huge section of Overtown in 1967. Source: "Miami, FL: Overtown," January 3, 2019, http://miamiplanning.weebly.com/freeway-history--overtown.html.

CHAPTER 1

an area that limited the city's economic growth, and they used data as evidence to make that point. Planning historians believe that the destruction of the neighborhood (figure 1.26) aligned with Miami businessmen's racial goals for their city, and according to several accounts, with their conviction that the African American presence would stifle the city's development.[67] This pattern of racist reasoning became standard practice during urban renewal and led to the destruction or marginalization of many more African American neighborhoods across the country, with data used as evidence to garner the necessary funds to make it possible.

Qualitative Analytics: The Call of Social Reformers

The failures of urban renewal policies, which often exacerbated problems in urban areas, caused some urban activists to fight the top-down planning methods that were being employed by technocratic planners. Jane Jacobs, an influential civic activist who became well known for her work in New York City in the 1960s, along with other civic actors such as Herbert J. Gans, led this charge.[68] Jacobs, who lived in New York's Greenwich Village, believed human-centered narratives were essential for understanding the economic and social needs of cities.[69] Jacobs fought against using data-driven policies as evidence for building highways because the results were so devastating to entire neighborhoods—and saw the removal of these neighborhoods as a complete failure to acknowledge the importance that their residents and networks played in the larger economy of the city.[70] Jacobs believed the public's voice should speak louder than data in the decision-making process.

It is interesting to note that neither Jacobs nor Gans argue against data *per se*, but rather against how data was typically collected and analyzed. Their focus was on understanding how the social connections of place created certain ties that strengthened the urban economy. Their interest leaned more toward qualitative data, including local stories and imagery. Unlike the data employed by the technocratic planner, this data included the perspective of the public because it was built by them.

Jane Jacobs's ideas about the city were often pitted against those of Robert Moses, New York City's most influential technocratic planner, who focused on making New York a "modern" city of highways and high-rises. Moses was able to deploy urban renewal strategies that effectively severed community ties, similar to what happened in Overtown. He did this using data as

1.27 "Mrs. Jane Jacobs, Chairman of the Comm. to Save the West Village, holds up documentary evidence at press conference at Lions Head Restaurant at Hudson & Charles Sts." Jacobs used here a different type of data—signatures of citizens. Source: Photo by Phil Stanziola, World Telegram & Sun, December 5, 1961, LC-USZ-62–137838, Library of Congress, https://www.loc.gov/pictures/item/2008677538/.

evidence. Jacobs, in contrast, came to represent a more qualitative planner who understood that the social networks of people are what make the city a great place to live, and that preserving those social networks, which help build and maintain community, were a legitimate focus of the urban planner's concern. For Jacobs, such social and support networks were often essential for low-income communities—and the modernist city as Moses envisioned posed a great threat to them.

Moses and Jacobs are often cited to represent two forms of city planning: the technocratic planner, who preferences the use of data to create evidence for modernist ideals, and the social planner, who preferences working to strengthen communities through building policies that support their needs and reinforce social ties. While the two are often juxtaposed, one using data and the other not, I would argue that both used data to create evidence for

1.28 Robert Moses, Park Commissioner of New York City, with a 1939 model of the proposed Battery Bridge. Moses is known for using data as evidence to send highways through poor communities. Source: Photo by C. M. Spieglitz, *World Telegram and Sun*, 1939, LC-USZ62-136079 Library of Congress, http://www.loc.gov/pictures/item/2006675178/.

their visions of the city, but they did so to different ends. In fact two famous pictures (figures 1.27 and 1.28) illustrate this eloquently: Jacobs is holding documents with citizens' signatures, while Moses stands in front of one of his urban renewal projects holding a report full of data analytics offering evidence for the development. Both are powerful documents, powerful pieces of data. Jacobs used her data to save a neighborhood from destruction, and Moses displaced whole communities to realize his modernist dream. Here we can see how using data for city planning can be a double-edged sword: it can be used to empower some and marginalize others.

With both sides armed with their own forms of data (qualitative versus quantitative), the debates between the technocratic planners and community advocates helped drive the development of public participation planning, which sought to narrow the gap between those who had control of information and decisions and those who did not. Shelly Arnstein produced seminal work on this topic to advocate for greater knowledge sharing between what she called the "power-holders" and the "powerless" in public participation.[71] Inherent to Arnstein's approach was the need to engage broader publics, including communities typically marginalized by public participation processes, because they do not have the resources and data to advocate for their needs.[72] Interactive education and community dialog became important tools to fill that gap. Similarly, James Glass, who writes about citizen participation in planning, believed that information exchange that involved education, building support, supplemental decision-making, and representational input could change power relationships within public engagement strategies.[73] John Forester, a leader in the development of the field of participatory planning, believed that through interactive dialogue, power could be transferred, as dialogue allows groups to present their perspectives and ideas and thereby educate each other on critical issues.[74] This early work was influential in my development of the Data Action method. The idea that engagement with data is one way to help overcome the oppression it might cause.

As opposition to the methods of technocratic planning rose in the 1960s, the urban theorist Kevin Lynch began to develop methods for transforming the map from a tool of control to one of public expression. His method, detailed in his book *The Image of the City*, involved asking community members to draw "mental maps" of their cities.[75] Behind his cognitive mapping exercise is the idea that the elements of the city most important to community members will be those details they remember and document on maps they draw from memory. Therefore, looked at collectively, these cognitive maps show what is important to the people who live there.

Maps: Good or Bad?

What are we to make of these numerous examples that show how maps and data have been used to marginalized populations? Should we avoid maps as a tool? Avoid data? The title of this book might provide a clue. Data can be used for action: it's how we work with data that determines its potential to help society. Ultimately these historical tales show that maps and data reflect the biases of the people who are using them as a policy tool. Booth's maps had the consequence of removing whole communities from the city. Bartholomew's maps were used as a tool to systematically and purposely segregate cities.

Later we see the federal government using the HOLC maps to foster further disinvestment in communities. In each case, these maps could have aided these communities rather than push them down. Sociologists such as W. E. B. Du Bois attempted to use the same type of maps as a counter-narrative to the prevailing idea that African American communities were full of "unemployed criminals." Each map has its own history, and each cartographer had an agenda. They all used data to support their ambitions—aspirations that often left those on the margins of society without the ability to fight back against oppressive policies.

To use data for action, we must interrogate both our intentions and the intentions of others. That is why it's so important that we show the maps we make to the people whose lives are included in the data that these maps rely on. Then we must ask: Does our analysis ring true to them, or does it reflect our own biases about the communities we seek to help?

The Technology of War: Data and Cybernetics

After World War II, we see a great deal of military technology applied to everything—and that included city planning. Planners shifted from being inspired by sociology to being inspired by the efficiencies in management developed during the war. In *Warfare to Welfare: Defense Intellectuals and Urban Problems in Cold War America*, Jennifer Light discusses how experts at governmental think tanks and aerospace companies in the 1940s, 1950s, and early 1960s found new work applying their ideas to cities in the 1960s and 1970s.[76] Cities were looking for new strategies, and these specialists were looking for new ways to apply their expertise, especially as government funding for the military was starting to decrease. The federal government began to fund research institutions with an expertise in military technology such as RAND, SDC, and SAGE and asked them to apply their techniques toward civilian use.[77] The Johnson Administration's "Great Society" used modeling projects developed by the Pentagon and RAND to support this shift toward cities.

City planners turned to everything from cybernetics and computer simulations to satellite reconnaissance.[78] According to Peter Hall, writing in 1989 for the *Journal of the American Planning Association*, for city planners during this time, "anything that could not be expressed in numbers was inherently suspect."[79] Data was central to these movements because it provided decisions with legitimacy, as the use of data was often equated with truth.[80] Los Angeles is a great example of a city that integrated some of these methodologies early on, developing a Community Analysis Bureau (established 1966) whose charter was to use data to make recommendations for everything from crime rates to unemployment and traffic.[81] Reports that came out of this bureau encouraged policies to increase the minimum wage, to help enroll children from the neediest families in preschools, and to establish affordable housing, among other things. But while the analysis proved useful, some argue that the work of the bureau was never truly integrated into the planning processes of Los Angeles; with no pathways to create planning action, it ultimately failed. Despite that failure, the bureau performed the data analytics necessary to secure important grant funding and turned to more policy-driven questions, unlike some of the models developed by enthusiasts of cybernetics.[82]

Cybernetics, championed at MIT by Norbert Wiener, is an interdisciplinary approach to examining how humans and machines communicate with and control each other. Its use for urban planning provided the ability to perform computational analysis that included theories of the city as a biological system and machine.[83] Jay Forrester, a researcher and professor at the MIT Sloan School of Management, applied cybernetics theory to modeling the dynamics of industry, including supply chains and resource flows.[84] After a chance meeting with John F. Collins, a former mayor of Boston, Forrester, who had no background in urban planning, set his sights on applying these ideas to cities, creating the Urban Systems Laboratory at MIT.

Forrester's experiment in modeling urban systems could be considered one of the biggest failures in using data analytics to understand, measure, and plan cities, and it helped to punctuate the problems of large-scale models to address planning problems. This failure stemmed from the fact that the model his lab developed, which was meant to be generalized to any city, made predictions that were completely out of line with the contemporary understanding of city planning. For example, it called for the clearance of slums to be replaced by luxury housing, and the inhabitants sent to public housing, which by 1969 had already proven to be an extremely problematic solution.[85] Some scholars attribute the failure of Forrester's model to its lack of focus on any one city—and therefore the variables could not be adjusted for any particular city—which made it hard to correlate the findings to real city behaviors.[86] Others noted the lack of understanding of the spatial dynamics of cities and the interdependence of the inner and outer city (suburbs).[87] Perhaps the mostly highly cited critique, *Requiem for Large Scale Models* (1973) by Douglass B. Lee Jr., argues that large-scale models, such as Forrester's, don't start with theory-based policy questions, thereby creating "empty-headed empiricism."[88] Ultimately Lee calls on urban modelers to make models more transparent and focused on specific policy questions rather than dealing with the city as an abstraction.

Inaccurate models, high costs, and the reinforcement of hierarchical power structures ultimately led many cities to give up urban modeling developed out of the military in the 1970s.[89] The critical-planning theorist John Forester (not to be confused with Jay Forrester) argued that little consideration was given to the fact that data interpretations were heavily biased as a result of bias from those generating and collecting the information.[90] The construction of models was often flawed because of the lack of appropriate data to develop them, the use of biased information, and the fact of not testing whether the data results fit people's experiences. A telling example of this is the 1970s-era model developed by the RAND Corporation for the New York City Fire Department to determine the efficiency of its network of firehouses.[91] Basing the model almost solely on response-time data, the analysis led to recommending the removal of several fire stations in the South Bronx. The data model did not consider additional data factors such as gridlock, politics, multiple simultaneous fires (and therefore the need for back-up units), and the socioeconomic status of the neighborhoods in question. This resulted in an overload on demands for the fire stations that remained in the Bronx, and it is estimated that fires displaced half a million people during this period. The RAND model did not *cause* the rampant fires in the Bronx during the 1970s, but it did incorrectly assess how the closures would affect the other firehouses in the network, thus placing undo stress on the system and displacing many people.

An Urban Planning Tool: Public Participation GIS

The advent of desktop computing in the 1980s and early 1990s meant that data analytics of the 1960s were more accessible to the public. This ultimately led to the development of geographic information systems (GIS), often referred to as visual databases, which were designed with the capability to capture, store, and link geographic data and allow users to perform interactive queries and manipulate the data to extract new insights. Although the GIS systems allowed for modeling, they also became a tool for managing city infrastructure—including detailed surveys of the cities performed by tax assessors to determine property values. The ill-fated models of the late 1960s and 1970s weighed heavily on the minds of early adopters of the software, who recognized that GIS could both marginalize and empower different populations depending on how it was used.[92] This led to a series of debates that began in November 1993, often referred to as the Friday Harbor meetings because of their location in the San Juan Islands of Washington State. During these debates, leaders in the GIS community discussed how it was difficult to use analysis from GIS to address social agendas because GIS software was often operated by those in a position of power.[93]

Ultimately these discussions led to the development of the field of Public Participation Geographic Information Systems (PPGIS), which attempts to address the power dynamics involved in spatial analysis and mapping by providing a methodology with which the public can more easily engage and learn about the data used to make decisions with GIS.[94] PPGIS projects are based on the premise that such engagement in data analysis and collection efforts can allow community groups to have more power in the political decision-making process.[95]

Since the development of PPGIS, governments and nongovernmental organizations (NGOs) have used it to share data with wider audiences and thereby influence national-level policies.[96] Community-based organizations have also used GIS within their own organizations to develop maps that represent their own perspectives and voices as they attempt to influence neighborhood-level planning initiatives.[97] Advocates for PPGIS believe these methods empower citizens by creating a process that makes it possible for the public to work with GIS systems and thereby critique the output, but still ensure that community interests are represented. GIS systems have now become an essential tool for planners, used for everything from evaluating environmental risk to managing electricity lines. Many open-source GIS software packages (including web-based mapping) are now available to truly put the power of modeling at the fingertips of urban planners as well as novice users.

The explosion of this software has created a new group of data enthusiasts who are not always aware of the risks of using data. This has produced a lot of "bad" or inaccurate analysis, not necessarily out of malice, but rather a lack of data literacy. For example, some maps floating around on the web often misrepresent basic census demographics, as data novices don't know that they have to normalize census data for population and areas. More problematic, because the work is represented as a map, is the fact that the general public often does not question its legitimacy, as maps are so often equated with truth. Data literacy is essential for these data enthusiasts as well as the general public.

Smart Cities and Big Data: Urban Modeling Rebooted

In the last decade there has been an upsurge of interest among private and public agencies in the uses of data for the design, planning, and management of cities and urban environments. Excitement around the potential of "big data" to change the way we see the world has created an enthusiasm for applying data to civic action and policy change. IBM developed its "Smart Cities" and "Smarter Planet" advertising campaigns in 2008 to promote the use of technology and data to analyze the problems of cities.[98] In 2010, IBM created a partnership with Rio de Janeiro to make it a model Smart City.[99] During that same year, Cisco launched its "Smart and Connected Communities" program, which is based on using data analysis and web-based interface programs to connect cities through technology.[100] Also in 2010, the *Economist* magazine developed a series of stories called the "Data Deluge," which explained the possibilities of using data to develop strategies for almost anything, including cities. The hype around these projects generated fascination in the popular press and media for what the analysis of big data can offer cities.[101] Microsoft launched its CityNext program, intending to draw from its worldwide network of technology experts to make cities better places.[102] In 2015, Google launched Sidewalk Labs, a grouping of "urban innovation organizations" working on various projects including data analytics with the US Department of Transportation and developing a "smart city" solution in Toronto.

Stephen Goldsmith and Susan Crawford document this interest in using data to drive government policies in their book *The Responsive City: Engaging Communities through Data-Smart Governance*, which illustrates how governments are harnessing the power of big data to increase efficiency in the delivery of city services. In the foreword, Michael Bloomberg, the former mayor of New York, explains why he thinks data is essential: "If you can't measure it, you can't manage it." He goes on to say that, "Harnessing and understanding data helped us to decide how to allocate resources more efficiently and effectively, which allowed us to improve the delivery of services from protecting children to fighting crime."[103] For Bloomberg, data analytics are essential for proper governance.

Urban planners have begun to shed their fear of data, which developed during the late 1960s largely in response to technocratic planning. In the fourth (and, to date, most recent) edition of *Introduction to Readings in Planning Theory,* Susan Feinstein and Scott Campbell insist that, like it or not, planners need "to deal with the coming flood of data." "Planners," they say, "need a larger conceptual world view to understand the ramifications of the digital revolution of the Internet, massive data storage and retrieval, and geographic information systems (GIS)." They argue that the questions planners have raised about reliance on "scarce, incomplete data" in the past may have been supplanted because data is now richer, and the processes to analyze and visualize it have been more fully developed.[104] Many planners embrace data, but we should continue to regard the intent of data analysts with a healthy dose of skepticism.

Open-source data access and online technology tools have allowed many other people, from opposition activists to NGOs to ordinary citizens—to create their own maps as well. GIS and computer mapping software are increasingly available, as are web-based mapping tools, allowing all sorts of non-traditional data visualizers to benefit from the potential these technologies offer to expose spatial patterns. The data used in these systems is also becoming more accessible, as cities and private data producers realize that opening data to the public means opening up possibilities—again, hopefully—for improving the cities, and for allowing everyone the chance to build upon and analyze the knowledge gained through public data collection. However, just because the data is more readily accessible doesn't mean everyone has the skills and knowledge needed to make use of it.

The misguided models of the 1960s and 1970s have largely been forgotten and replaced by an excitement for harnessing the power of big data for cities. Large corporations such as IBM, Cisco, and Microsoft, which can make a profit by applying their computing power to big data analytics in cities. Yet these analysts, who are often untrained in urban planning or related fields, are not always familiar with how data models have been used inappropriately in the past. It is therefore up to us to educate them to use data ethically and responsibly.

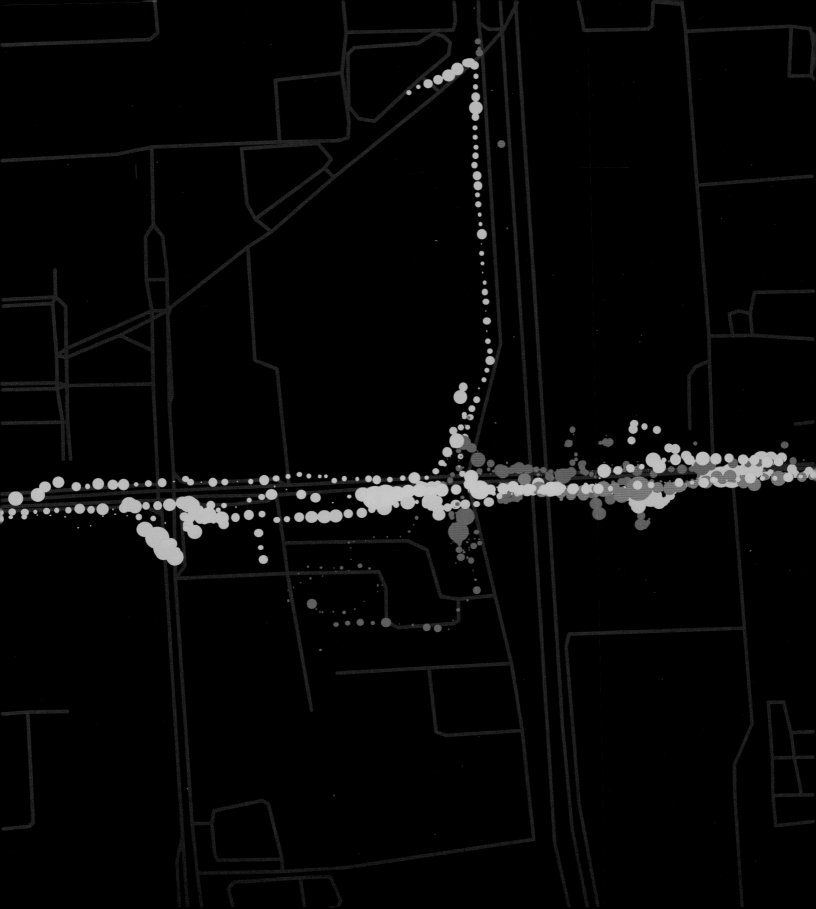

BUILD IT! DATA IS NEVER RAW, IT'S COLLECTED

We have the sixteenth-century philosopher Sir Francis Bacon to thank for the familiar idiom "knowledge is power," and one of the fundamental ways we gain knowledge by is collecting data. When it comes to cities and their systems, data collection has traditionally been designed by those with the resources to collect the information, and thus has largely been the activity of governments, corporations, and institutions. Now anyone can collect data with a little training because innovations in digital technology are turning everything from mobile phones to dog collars into data collection devices. Often referred to as participatory sensing projects, or crowdsourcing,[1] these citizen-based efforts in data collection fill gaps in knowledge by equipping people with the ability to collect raw data that was previously unattainable and then to process that data so it can be visualized, analyzed, or otherwise transformed. But perhaps more importantly these projects help build communities that together gather the evidence to both fight for and understand their rights. Participatory sensing projects also teach data literacy and give communities control over their narrative. Beyond creating the data itself, they are Data Action tools that provide a critical way of addressing power imbalances.

In the last decade leveraging big data has become big business, and tech companies want a piece of the action. But as analysis and collection of data about cities are further commoditized, we run the risk of excluding communities on the margins because their data is missing or not reflected in the analytics that corporations develop. In the introduction I mention the International Data Corporation (IDC) estimate, that in 2025 the world will create and replicate 175ZB of data: that astonishing amount represents more than a tenfold increase from the amount of data created in 2015.[2] And it's even more mind-boggling to consider that a single zettabyte is roughly equal to 250 billion DVDs. Private companies own the majority of this data, and it is largely inaccessible. The only way for some communities to counter prevailing narratives and work for change is by *building their own data* where governments and institutions have failed to do so.

Participatory Data Collection: The Act Itself Is the Tool

One method we can use to build our own data sets is to collect data with the participation of broader, interested, self-selected communities. An emphasis on such participatory and collaborative planning processes emerged in the mid-1980s; it emphasized using community data collection as a way to counter the data collected, controlled, and presented—and especially ignored—by governments and institutions.[3] This ideological shift, often referred to as participatory action research (PAR), has its roots in the theory of critical pedagogy that Paulo Freire outlined in his 1968 book *Pedagogy of the Oppressed*.[4] Fundamental to Freire's theory was the idea that education could help liberate the oppressed because they could more actively engage in a conversation with those in power.[5] Freire believed that poor and exploited peoples must construct their own narratives by collecting data on their own reality, and thus be co-creators of knowledge.[6] For others the movement toward more community driven planning processes was a backlash to government analytics projects from the 1950s, 1960s and early 1970s, which set policy agendas that often left those on the margins underrepresented, causing large social divides in the development of cities.[7] Whatever the reason for the shift, planners began to develop community-led data projects that have been referred to using numerous and varied terms, including counter-mapping, community mapping, participatory mapping, critical cartography, ethno-mapping, participatory GIS, Volunteered Geographic Information (VGI).[8]

Participatory sensing is the mostly widely used term to describe community involvement in data collection; it first appeared in an article by Jeffrey A. Burke and colleagues, where the authors describe the term as using "deployed mobile devices to form interactive, participatory sensor networks that enable public and professional users to gather, analyze and share local knowledge."[9] Many early participatory sensing projects made use of data in mobile phones, such as GPS data for location; others focused on using platforms available through mobile devices to send information amassed from sensors, which were placed elsewhere, back to centralized databases.[10] One project created a platform based on SMS (short-message service, or text messaging) to send air quality data from Environmental Protection Agency (EPA) sensors to users who texted requests for it; the responses provided real-time information on air quality.[11] Another project originating from Nanyang Technological University in Singapore created a system for Android phones that used signals transmitted between smartphones and cell phone towers to geolocate users and buses; it allowed users to submit and query bus arrival times based on their identified locations.[12] For the most part, participatory sensing projects aim to maximize the capabilities of mobile devices, reaching beyond their primary uses for voice and text communication to also collect data.[13]

What's interesting about these participatory data collection methods is that accuracy does not always matter. The data purist may find the lack of accuracy unthinkable, but because participatory sensing projects are often used to inform and teach citizens about an issue, the act of collecting the data becomes the primary learning tool. Any inaccuracies unearthed in the process are another opportunity to teach participants to be skeptical of any data they are presented with in the future. For the designer of participatory processes, planning the project to engage a diverse range of stakeholders and then maintaining their cooperation are as important as the data collection itself.

More and more people can now use data analysis to advocate for government policy or reform because sensor networks have become ubiquitous. Today, mobile phones come embedded with an array of sensors—from GPS, cameras, microphones, accelerometers, and more. Social media apps also help facilitate data collection projects through the sheer scale of their reach. Facebook, Twitter, and Foursquare reach vast numbers of people whose opinions can be harvested by place and time. Although there is and ought to be much concern about how this data is used by governments and private corporations, these ubiquitous data recorders can just as easily be leveraged by the public to answer questions important to society.

I discuss the ethical concerns of using data collected by private companies in more detail in chapters 3 and 5, but there are still ethical concerns to consider when members of the public collect data. When using data for action, the people affected by the data must be aware of the project and should also be able to critique the results. The public, like anyone else, collects data from its perspective and can intentionally or unintentionally use that data unjustly. It is essential that novice data collectors hold themselves to the ethical standards established by Data Action.

The impulse to volunteer information captured by mobile sensors is the basis for the term *volunteered geographic information* (VGI), first introduced in Michael Goodchild's seminal piece, in which he describes VGI as "a special case of the more general Web phenomenon of user-generated content."[14] VGI has largely been used to describe geographic data that is contributed through the web or mobile technologies. Though the term VGI seems largely relegated to the work of academic geographers, it was developed to highlight that a community of people exists who are interested in telling us *what* they are doing and *where* they are doing it, and if we can tap into that community we can use it to understand new social-spatial phenomenon.

Crowdsourcing a Global Data Set

A well-known example of VGI (or community mapping) is Open-StreetMap. By allowing participants anywhere to upload GPS traces, OpenStreetMap created and provides a worldwide basemap that is free for anyone to use. The basemap includes roads, parks, and major points of interests, and in some locales it even includes building outlines. OpenStreetMap was developed in 2004 by Steve Coast, then a PhD student at University College London, in response to the control of basic map data in the United Kingdom. At the time, the UK Ordnance Survey, the government's mapping agency, charged a hefty sum to acquire and use map data that it collected and maintained. This meant it was difficult to obtain basic census, postal code, streets, parks, and building data. Coast envisioned creating a Wikipedia for maps data by allowing any novice map hobbyist to contribute to his platform using verified data from handheld GPS units. Since its inception, OpenStreetMap has had close to two million contributors.[15]

Coast's OpenStreetMap was meant to free data he believed should be accessible to the public. Before OpenStreetMap, if you wanted base data for anywhere in the world, you would have to purchase it from two big companies, Tele Atlas and Navteq, and the data did not come cheap. Google

Maps used a combination of Navteq and Tele Atlas data during its first few years, and the costs for licensing this data rose so high that it became more cost-effective for Google to develop its own worldwide basemap; you may have seen a Google Maps car, with a camera mounted to its roof, driving through your neighborhood.

Steve Coast's frustration with getting his hands on affordable data evolved into a global phenomenon, and in 2006 the OpenStreetMap Foundation was started with the express mission of collecting and uploading openly available map data everywhere. Map nerds in cities across the world gathered at "map parties" promoted by the foundation's volunteer members, where they would learn how to use handheld GPS units and set out to map unknown territories. Map parties often simply involved sitting around a table in front of computers, pizza boxes nearby, tracing roads from satellite imagery using an OpenStreetMap tool. Map parties created a place for map enthusiasts to socialize and share information on the latest mapping technologies, and resulted in a strong open-source mapping community. Attending map parties was a way to learn from peers while generating data under the ethos of open-source principles and practices.

In 2007 OpenStreetMap held its first "State of the Map" conference, which has since become one of the main gatherings for the open-source map community.[16] Recently at this conference the Red Cross, USAID, and a number of other humanitarian organizations demonstrated techniques for building data for disaster relief using OpenStreetMap tools. For example, the Red Cross developed OpenMapKit, a tool that allows its staff to collect data in remote areas where limited internet and cell networks make accessing and uploading data to OpenStreetMap's database difficult. OpenMapKit solves this problem by making it possible to save the data on surveyors' mobile phones until, once in reach of cell networks, the data can be easily uploaded. The open-source map community that OpenStreetMap developed has helped drive innovation, particularly in applications—such as those designed to speed disaster response—where crowd participation is essential.[17]

OpenStreetMap is probably the most successful participatory-mapping project; it is known for its scope and relative accuracy and, importantly, has filled a gap in government information. Once logged in to OpenStreetMap, anyone can find functional basemaps of countless places around the world, an impressive feat given that most of this data is gathered by the public: for many cities the only existing data set on its urban infrastructure is found on OpenStreetMap. OpenStreetMap illustrates how building mapping data brings together many people in common cause to put information that matters to each of them into public hands; this open creation and exchange of information are the first collective steps to changing the status quo and using data to change society. OpenStreetMap is a great example of how Data Action sets out to create missing data needed for policy. Now I will take you across the world to show how the same method can free data in a place where it is tightly controlled by the government.

Exposing Beijing's Air Quality: DIY Data Collection

Data is powerful, and even more so in places such as China that tightly control information. It is no surprise that on the eve of the 2008 Beijing Olympics, no publicly available data on air quality was available. With the world's top Olympians set to descend on the city, there were serious concerns for the athlete's health and ability to compete. The first example of using data for action in this book tells the story of how my research team became the only source of air quality data during the Olympics and how this data reached a world audience through the international press. It ultimately shows how collecting data outside government can bring policy issues to the public—creating pressure for policy change. Perhaps more importantly, this case study provides an example of how anyone might leverage cheap sensors attached to mobile phones to *build* data where it does not exist, and then use it to start a public debate about the environment.

China's struggle with air quality issues was not news. In the months leading up to the 2008 Olympics, one of the biggest concerns was whether the air in Beijing would be harmful to athletes and visitors alike. The *Washington Post*'s Maureen Fan reported that many visitors to the city were struck by the city's gray skies, and they wondered, "How this city can possibly be ready to host [the Olympics] in less than ten months?"[18] The *Guardian* reported that "more than a million cars were taken off the roads for the four-day test period, but there was no improvement in the air quality."[19] The *New York Times* summed up the situation evocatively when correspondent Jim Yardley wrote, "Beijing is like an athlete trying to get into shape by walking on a treadmill yet eating double cheeseburgers at the same time."[20] The reports from Beijing appeared to suggest that improving the quality of the air was impossible.

The Chinese government assured the international community that air quality would be significantly improved by the time of the Olympics, and that it would enact certain policies to make that happen. There were to be widespread factory shutdowns. Industrial businesses were ordered to suspend operations in Tangshan, one of China's busiest steel centers, about ninety miles east of Beijing. The Chinese government also wanted to address air quality on congested roadways and removed over 3.3 million cars from the streets by mandating alternative-day traffic regulations for vehicles with even and odd license plate numbers. Just days before the Olympics, the *Washington Post* reported that an additional 220 factory closures were ordered in the city of Tianjin and in Hebei province as air quality did not seem to be improving there. Beijing also extended measures to include a ban on construction during the Olympics.[21]

Many observers, including the International Olympics Committee, were anxious to see if Beijing's policies would meaningfully decrease their problematic air pollution. I recalled my last visit to China where, as an avid runner, I could hardly make it around the block because the air quality was so poor. No one could see the blue in the sky. Like other visitors, I, too, thought, "How is Beijing going to improve its air quality in time for the Olympics?" And then, "How can I use the Olympics—when eyes around the world would be focused on Beijing—to expose air quality conditions in China?"

I attended a presentation by Patrick Kinney, from Columbia University's School of Public Health, about giving a group of middle-school students air quality sensors to put in their backpacks so they could understand how the urban environment affected their asthma.[22] I was inspired by my colleague, who used the data collected from these sensors to determine if the students' daily movements—and the air quality at those locations—triggered an asthma attack. Could similar mobile sensors, I wondered, be used during the Olympics?

After researching mobile sensors, I realized that it would be possible to obtain a relatively cheap mobile sensor to test air quality in real time. Early mobile sensors were sent to a lab for data interpretation, and it could take months for the results to be returned—this is how my Columbia colleagues worked. New sensors, however, allow measurements to be taken in the field, which means no waiting for days or weeks to get the results. Given the potential of a quick response turnaround with these new sensors, we decided to follow athletes, such as track and field specialists, whose lungs would be most at risk from excessive pollution. Logistics, too, needed to be addressed: How would we gain access to the appropriate Olympic venues? Would tickets to the Olympics be necessary? Could we linger outside the venues even if we didn't have tickets?

I mentioned our plan to a friend, a journalist for the Associated Press (AP), who enthusiastically offered to propose the idea. The AP was excited to report on the air quality during the Olympics, but it had never collected its own data before. Instead it had gravitated to projects where the data was already produced by a research organization, preferring to put the scientific burden on the experts rather than be on the line for data accuracy. Nevertheless, the idea traveled up the chain of command. The AP ultimately asked my research lab to develop a proposal to work with its reporters. A great partnership formed; the AP would get the data they needed, and we would both be able to use it to further a public dialogue about air quality conditions in China.

Unfortunately, the AP's interest in the project waxed and waned. When air quality was a hot news topic, because an athlete or government mentioned concern, the AP would start moving forward with the proposal. But when the news cycle moved on to something else, so did the AP's attention. The back-and-forth continued for close to four months; with only two months left before the opening ceremony we still did not have a signed contract with the AP to undertake the project.

Yet through all the ups and downs I believed the project would be a groundbreaking example of how employing mobile sensors could start a conversation about poor air quality in China. So, I decided to build my idea without funding from the AP. I remembered reading *The Revolution Will Not Be Funded*,[23] which inspired me to take the risk of building an air quality sensor platform that would connect to cell phone GPS units and would be given to all the Associated Press reporters. I knew it was a good idea and figured the AP would want the sensors—at the last minute. So I pulled together some extra resources, and, sure enough, just weeks before the AP team was set to travel to Beijing, another news story broke about the air pollution, but little data existed to back up anecdotal reports. The AP needed the data my sensors provided, so the agency finally signed the contract. The sensors were ready, and, with only a few days to spare, they were placed into the AP equipment kits that were being shipped to the Olympics's press offices in Beijing. See figure 2.1 for photos of the AP team (and their kits) in action.

2.1 This photomontage shows AP reporters as they were measuring particulate matter and carbon monoxide during the 2008 Olympics in Beijing. Source: Photos by Sarah Williams and J. Cressica Brazier.

The Sensors

For the Beijing Air Tracks—as the project came to be called—air pollution sensors needed to be highly mobile and portable so that AP reporters could carry them effortlessly. But small sensors can only measure a limited number of pollutants. We chose to measure pollutants that would tell us the most about not only the environmental conditions that would affect the athletes, but also the Chinese government policies designed to improve those conditions. We therefore focused on collecting fine particulate matter and carbon monoxide.

Fine particulate matter, or PM, is a broad name given to particles of liquids and solids that pollute the air. We inhale PM into our lungs where it remains, causing respiratory and other health problems. PM was critical for telling the

2.2 Tiananmen Square at noon, just one month before the official start day of the 2008 Olympics. The Beijing Air Quality index was marked as unacceptable that day, and the pollution can be clearly seen in the image, as the gates of the Forbidden City are shrouded in a hazy "smog soup." Source: Photo by J. Cressica Brazier.

2.3 This map shows data traces measured by the CO sensor of an AP reporter near the hotel where he was staying in Beijing, prior to the start of the Olympics: the larger the circle, the higher the CO concentration. Each color represents a different day on which he tested the measurement. Source: Image by Sarah Williams.

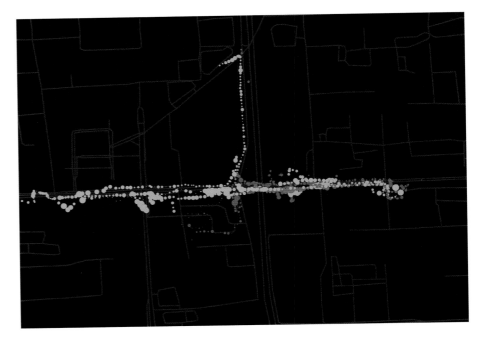

story of air pollution in Beijing because this category of pollutant would have the largest negative effect on athlete performance, and because it was related most directly to the Chinese mandate for widespread factory shutdowns.[24] Chinese factories, which are largely powered by coal, are arguably the largest contributors to PM pollution, which can travel long distances by air, and thus affect areas geographically far removed from the site of production.[25] We chose carbon monoxide (CO), typically produced by vehicular traffic, because we felt it was the pollutant most readily understood by the public. Real-time CO sensors are also more compact and cost-effective than similar sensors for most other air pollutants.[26]

The fast-paced nature of Olympics news meant we needed air quality sensors that could measure data in real time. Real-time sensors are generally less accurate than static sensors, which are usually left on rooftops and can better handle sensing anomalies caused by wind and weather conditions. While our mobile sensors might not deliver the same accuracy level as hyper-calibrated static devices, we believed that sharing our findings, with full disclosure of possible errors, was better than no action at all.[27] The team also knew that we might find pollution levels to be so high that if we undercounted the data, the sensor recordings would still show levels dangerously above World Health Organization standards (see figure 2.2).[28] Figure 2.3 shows CO levels measured by an AP reporter.

Our real-time sensors were connected to GPS receivers, and the sensor data was therefore tagged with a geographic location. Once each sensor measurement was tagged with a latitude and longitude, the data would be relayed to the main press center via the cell phones.

Measuring and Visualizing

Six weeks before the Olympics started, we gave the sensors to five AP reporters and set out to collect particulate matter and CO levels at well-known locations throughout Beijing in order to test how well the sensors performed (figure 2.4). The team focused on making daily recordings in the Olympic Park, Tiananmen Square, and the Temple of Heaven to have a baseline of air quality before the Olympics began. Just days before the opening ceremony, Chinese government officials questioned two AP photographers using the air quality devices about photographs they had taken. The government didn't notice the air quality sensors, but the surveillance of these reporters was cause for concern. The AP took all but one reporter off the project, fearing that the Chinese government might detain them and thereby prevent them from obtaining important photos of Olympics events.

The team focused the project's output on two data visualization stories that AP member newspapers across the world could purchase. One would show PM levels in Olympic Park; another would show CO levels on the marathon route before and after the Chinese enforced a policy to remove half the cars on the roadways. The Olympic Park PM-data visualization focused on presenting PM data recordings contextualized by the World Health Organization (WHO) standards. The visualization also used photographs to capture the atmospheric conditions, which often looked like a hazy soup on

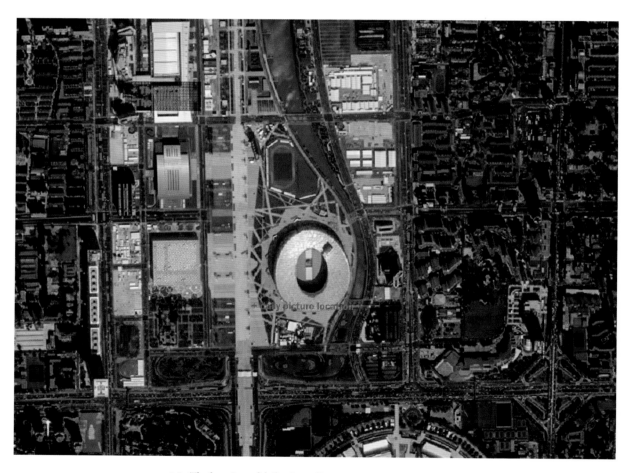

2.4 The location of daily air quality measurements AP reporters took on the Olympic Green. Source: Image by Sarah Williams.

poor air quality days. The daily routine of one AP reporter included taking a photograph of the Bird's Nest Stadium from the same vantage point at mid-day, while running the air monitor attached to his backpack. That reporter was also collecting data for the interactive graphic. The AP's graphics piece used these daily photographs of the Bird's Nest as a navigation device: when users clicked on a photograph, the PM recordings would appear for that day (figure 2.5). The graphic also compared PM numbers with average readings in London and New York for the same day. It showed that the level of particulate matter in Beijing's air was ten to twenty times higher than what it was in New York City and London, and that on nine out of the sixteen days

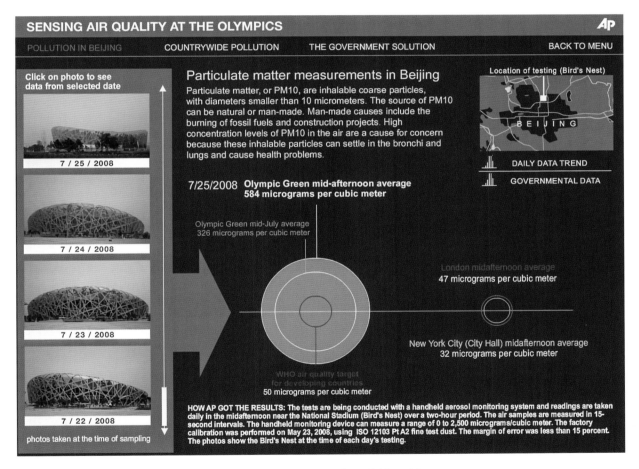

2.5 Interactive graphic of particulate matter on the Olympic Green made available to AP member institutions. Source: Image created by Sarah Williams and Siemond Chan for the Associated Press, 2008.

2.6 The graph of raw data from the particulate matter sensor used on the Olympic Green by the AP during the Olympics shows Beijing in comparison to New York and London for the same days. Beijing has levels often ten times higher than New York and high above World Health Organization standards. Source: Image by Sarah Williams.

of the Olympic Games, the particulate matter levels did not meet the World Health Organization's standards (figure 2.6).

The research team also created an infographic that showed geo-registered CO readings taken along the Olympic marathon route before and after the Chinese government ordered 3.3 million cars removed from the roadway. The infographic, developed in coordination with the *New York Times*, focused on CO levels because the story was more closely associated with vehicle exhaust (figure 2.7). My research team and the *New York Times* compared the figures in Beijing to CO measurements obtained along New York City's marathon route on the same day.[29] Analysis of the data conveyed through the visualizations showed that in many areas along the Olympic route, CO levels

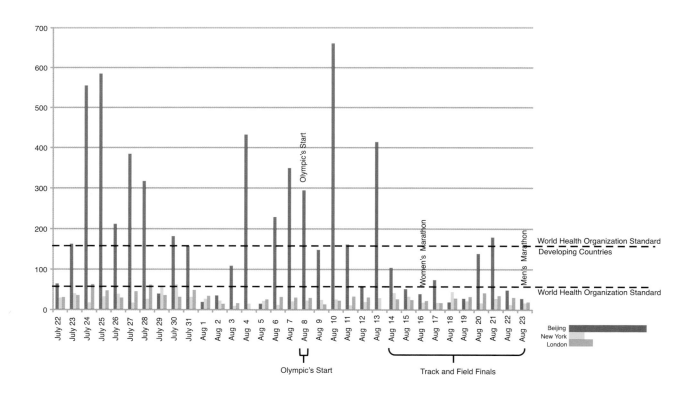

OLYMPICS
BEIJING '08

'I just ran the first 50 meters, then I looked around
to make sure I was safe and I shut it off.'

USAIN BOLT of Jamaica, who coasted through early-round qualifying in the 100 meters

Beijing Univ.

FOURTH RING RD.

■ **Finish**
Bird's Nest
National Stadium

A

THIRD RING RD.

These two
stretches had
some of the
highest CO
readings.

SECOND RING RD.

Purple
Bamboo
Park

Beijing
Zoo

THIRD RING RD.

Forbidden
City

Start
Tiananmen
Square

Tiantan
Park

BEIJING BEFORE RESTRICTIONS
Wednesday, July 9
Olympic marathon route
Avg. CO level: 5.9 p.p.m.

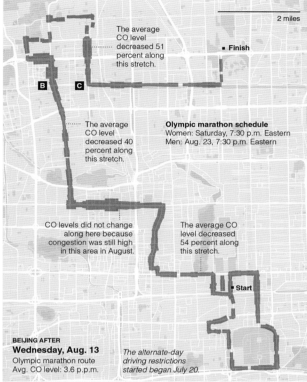

The average
CO level
decreased 51
percent along
this stretch.

■ **Finish**

B   **C**

2 miles

The average
CO level
decreased 40
percent along
this stretch.

Olympic marathon schedule
Women: Saturday, 7:30 p.m. Eastern
Men: Aug. 23, 7:30 p.m. Eastern

CO levels did not change
along here because
congestion was still high
in this area in August.

The average CO
level decreased
54 percent along
this stretch.

• **Start**

BEIJING AFTER
Wednesday, Aug. 13
Olympic marathon route
Avg. CO level: 3.6 p.p.m.

*The alternate-day
driving restrictions
started began July 20.*

Idling
buses
caused
spikes
along Fifth
Avenue.

QUEENS

Finish
Central
Park

MANHATTAN

Construction
machinery
operating here
caused a
spike in the
CO levels.

BROOKLYN

2 miles

STATEN
ISLAND

Start

NEW YORK CITY
Wed., Aug. 13
2008 marathon route
Avg. CO level: 3.2 p.p.m.

Measured Improvement

The Chinese government's effort to improve
Beijing's air quality appears to have worked to
reduce carbon monoxide emissions, based on
preliminary data collected by Columbia
University's Spatial Information Design Lab. In
a series of measurements taken along the
Olympic marathon route, carbon monoxide
levels dropped an average of 40 percent after
the city put in alternate-day driving restrictions
in which motorists could drive only on odd or
even days. Unfortunately, other components of
the city's air pollution, like particulate matter,
are still at high levels, according to Sarah
Williams, director of the lab.

A

*One of the main reasons for Beijing's
high carbon monoxide levels is the large
number of cars idling in traffic.*

B

*This neighborhood still registered high
levels after the restrictions because
there was still traffic congestion.*

C

*Most of the improved areas, like
this one, had dramatically fewer
cars than before the restrictions.*

KEY TO THE MAPS
Carbon monoxide levels

1   5   10   15   20
PARTS PER MILLION

Levels were measured around
midday. In a few locations, the
routes could not be exactly followed
because of obstructions, like a fire
in one case.

*Source: Sarah Williams and Cressica
Brazier, Spatial Information Design Lab*

ARCHIE TSE/THE NEW YORK TIMES

2.7 Carbon monoxide measurements along the Olympic marathon route before and after the Beijing
government restricted 3.3 million cars from the roadways. The results show air quality conditions
similar to what would be seen along New York City's Marathon route. Source: Image provided to the
New York Times by Archie Tse and Sarah Williams.

were reduced by almost half, and levels overall were comparable to what would be found in New York City. The marathon route was a compelling way to tell the story of air quality because it was an event that readers could connect with, plus the news was, in some senses, "good": CO measurements revealed that car reductions enforced by the Beijing government did help CO emission in some parts of Beijing.

During the Olympics, official Xinhua news articles referred to poor air quality conditions as "fog, not smog." The government's official air pollution index gave numbers ranging from 1 to 500, which made the index hard to interpret. How, for instance, did they rate varying levels of various pollutants? The lack of transparency in the creation of the data index often resulted in government officials having to explain to reporters what the index numbers meant.[30] Measuring the air quality independently allowed the press to discuss the topic with verifiable numbers and provide a fact-based counter-opinion to the smog-as-fog argument. Yet measuring the data alone did not change the conversation; creating compelling graphics for a large, captivated Olympic audience allowed us to bring the story of air quality in Beijing to people all over the world (see figure 2.8, for images shown at NYU).

The Beijing Air Tracks project illustrates that collecting your own data can set the record straight. The *New York Times* and the Associated Press were able to validate their stories on air pollution with data. They were able to counter the Chinese government's argument that what they were experiencing was fog, not smog, as they had hard data to prove the haze over Beijing's sky was caused by high levels of particulate matter. The project also educated the public about what air quality levels meant by allowing readers to compare levels in Beijing to other locations they may have been more familiar with. Perhaps most importantly, it set an example of how, on a relatively modest budget of about $15,000, devices can be made that can have a global impact. After the Olympics, the Beijing government continued its policy of removing cars from roadways because CO levels improved so significantly during the Olympics.

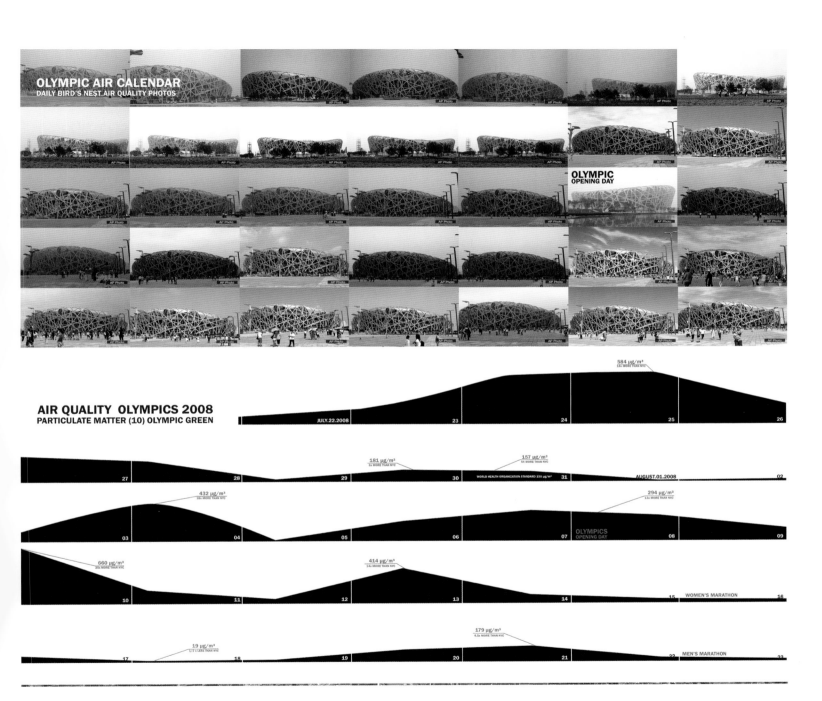

2.8 This visualization presented at New York University's Global Design Exposition shows pollution at the Bird's Nest Stadium in Beijing using photos and a graph of particulate matter data. Source: Images by Sarah Williams.

Data Collection: Cross-checking and Correcting the Record

Participatory sensing projects can be a way to crosscheck existing government data (or to supplement it) in order to make the public more aware of a controversial issue. This was the case in Beijing, where the government's official air quality index combined the data in a way that had made it hard to extract the measurement of a single toxin. Perhaps more importantly, there was ample suspicion that the Chinese government would not release accurate air quality measures in the first place. To create a compelling counter-narrative to dubious government claims, one needs data to be in the hands of the public.

Many environmental data collection projects attempt to fill data gaps this way. These projects are not necessarily trying to create a comprehensive data set (although some do), but rather provide some indication of environmental conditions to counter or supplement official governmental information. These projects provide solid evidence of a problem, which makes it easier to start a conversation that might lead to change. The Citizen Sense project, for example, which began at the Department of Sociology at Goldsmiths, University of London, and was run by Jennifer Gabrys, uses open-source hardware sensors to measure PM and byproduct toxins released by fracking in Pennsylvania (nitrates, sulfates, organic chemicals, metals, oil). Many residents in Pennsylvania shale country believe that compressor stations—facilities that pressurize natural gas extracted from shale for transport through pipelines—were polluting the air and causing health problems. In 2014, the Citizen Sense team suggested that people living near these compressor stations use a sensor called Speck (which is part of a sensor package they call FrackBox) to measure the quality of the air in and surrounding their homes.

One of the participants in the study, Rebecca Roter, found elevated levels of PM 2.5 micrometers over a seven-month period and sent the data to the EPA. The EPA responded by putting a higher-quality sensor in her area for an eighteen-day period and found her claims to be true. The EPA responded further by increasing its official network of environmental sensors in the region, which allowed the agency to monitor fracking compressor stations more efficiently. This in turn made it easier for them to enforce standards because based on the data they could now fine fracking stations if the levels did not comply with EPA rules. Rotor's sensors were not high-tech, but deployed cheaply they had significant impact.[31] Gabrys points out that having a highly accurate network of sensors does not always matter, but being able to wield data that showed heightened levels was "just good enough" to get the government to establish better monitoring experiences that allowed for stricter enforcement—illustrating that participatory sensing can have an impact even if the data is imperfect.[32]

Governments and private institutions often worry about the release of data, as they may not have control over the conversation that results; this was the case following the 2010 British Petroleum (BP) oil disaster. In April 2010 an oil pipeline burst, spraying unprecedented amounts of oil into the Gulf of Mexico from the waters off New Orleans, Louisiana, to Pensacola, Florida. BP appeared very reluctant to share information about the extent and spread of the spill. According to the *New York Times,* "three weeks passed, for instance, from the time the *Deepwater Horizon* oil rig exploded on April 20 and the first images of oil gushing from an underwater pipe were released by BP."[33] Congressman Edward J. Markey (D-Massachusetts), who fought to have BP release the video, commented that the company "was not used to transparency. It was not used to having public scrutiny of what it did."[34] Many journalists hired private planes to take pictures that would allow them to map out the damage, but at times they were not able to gain access to the spill location because they were told the air space above it was restricted.

Lack of access to data about the spill drove Public Lab, a group of environmental activists, to create a makeshift DIY toolkit for aerial photography out of helium balloons, kites, and inexpensive digital cameras (figure 2.9). Photographs of the spill taken daily from great heights were then knitted together in a program called MapKnitter, created by activists themselves. Their balloons were flown at a height of roughly one- to two-thousand feet, resulting in images of a higher resolution than Google Maps offered. These images could be used to identify more detailed views of environmental damage.

This Public Lab project is an example of "counter-mapping," which refers to efforts to generate maps that are precisely "against the dominant power structures, [in order] to further seeming progressive goals."[35] Counter-mapping has been used most often in natural resource and environmental protection efforts: the community helps to identify land and resources, and uses the evidence to counter claims made by the government. Nancy Lee Peluso first coined the term in 1995 to describe her work in Kalimantan, Indonesia, where she used the process of mapping with the indigenous

2.9 Image of the balloon mapping kit one can receive in the mail from Public Lab.[38]

population to identify their claim on forest resources. The maps helped to challenge those developed by a government that was undermining and minimizing the claims of the area's indigenous population.

The motivation for the Gulf of Mexico oil-spill mapping project was straightforward: the team wanted evidence regarding the extent of the spill to be developed independently of BP or the government.[36] Each member of the team shares a different story about how the collaboration started. Jeff Warren, an MIT student who began developing the balloon technology for his master's thesis before the oil spill, says that he contacted Shannon Dosemagen, who was part of the Louisiana Bucket Brigade already trying to document the spill, to see if she would be interested in using his new technology. In an article written about the project, Dosemagen recalls that she and Warren assembled a diverse group with different expertise: "it started with seven of us—an MIT student, a geographer, a biologist, an artist, a designer, a health scientist," and Dosemagen herself, an anthropologist.[37] This group trained hundreds of volunteers across New Orleans to use their kit and collected over 100,000 images. In a situation where there was no advantage for BP or the government to release data to the public, the citizens of the Gulf themselves gathered it, enabling them to build a data-informed narrative about the spill that they could trust. This project illustrates how using data for action does more than build data; it builds communities through collaborative data creation. These communities also use their networks to send the insight they find in the data to broad groups.

According to Dosemagen, the team's work also made an impact by exposing conditions on the ground to wider audiences. "News outlets including the *New York Times*, the *Boston Globe*, ABC, and CNN used the images in their stories. Lawyers requested copies for litigation. Google Earth posted the group's maps on their websites."[39]

Dosemagen referred to this practice as "community science." Instead of answering a researcher's questions, "Community science empowers ordinary people to ask their own scientific questions and follow through with data collection and analysis. The process is as much about science as it is about education, community building, and curiosity."[40]

The collaboration that started over the oil spill launched the nonprofit Public Lab, which refers to itself on its website's homepage (publiclab.org) as a community to investigate environmental concerns by developing do-it-yourself (DIY) techniques and tools. Public Lab's Balloon and Kite Mapping tools kits are now sold on Amazon.com, and community activists inspired by the lab's ideas have mapped everything from the dangers of urban slums in Nairobi (figures 2.10 and 2.11)[41] to pollution in the Gowanus Canal in New York City (my own first experience with the technology). In all of the stories of balloon-mapping projects, what strikes me most is how the communities work together to generate the data and more importantly bring their findings to broader groups through the novel methods of their data collection. The tactile nature of the project creates a shared experience—creating community around a civic interest—markedly different from buying a satellite image. Building data does more than make data, and that is very much in the spirit of the Data Action method.

2.10 Participants engage in a balloon-mapping project. Source: S. Ahmed et al. 2015. Cooking Up a Storm: Community-Led Mapping and Advocacy with Food Vendors in Nairobi's Informal Settlements. IIED Working Paper.

2.11 People identify buildings in a balloon-mapping project. Source: S. Ahmed et al. 2015. Cooking Up a Storm: Community-Led Mapping and Advocacy with Food Vendors in Nairobi's Informal Settlements. IIED Working Paper.

Crowdsourced Data: Why Governments Need It

We expect governments to have the data they need to make informed decisions—whether or not they steward it well—but that's not always the case. Crowdsourcing techniques can help governments to generate that data. An exemplary project of this sort is Safecast. It grew out of a desire to map radiation levels after the Fukushima-Daiichi nuclear disaster in Japan in 2011 and is now a global project working to empower people with the use of data. The Fukushima-Daiichi nuclear power plant suffered major damage from an earthquake (rated as magnitude 9.0) and a subsequent tsunami. Several reactors on site were permanently damaged, and all have since been decommissioned. While the government was monitoring radiation in and around the site, its sensors were not sufficiently spread out geographically, which made it difficult to know to what extent and to what degree communities near the plant were affected. According to several reports, the power plant operators and several governmental organizations, including the Japanese Atomic Energy Agency, failed to share information with the public.[42]

The Safecast team developed *bGeigie sensors*, short for "Bento Geiger Counter," a name they were given because the sensors looked like Japanese lunch boxes.[43] These sensors, developed soon after the disaster, comprised existing technologies including a Geiger sensor and GPS attached to an SD card device to save the data. The team also added the means for people to upload data from the sensor to a collective database. Ultimately, the team developed a more streamlined sensor, which could be bought as a DIY kit from the KitHub website.[44] Rather than going to test the sites themselves, the Safecast team decided to employ *crowdsourcing* techniques to collect data from all over Japan.

Crowdsourcing is a term that describes a spectrum of initiatives involving (sometimes) massive numbers of people participating in task-based problem solving. Jeff Howe codified the term's more contemporary meaning in a 2006 article in *Wired* titled "The Rise of Crowdsourcing," using the term to describe the strategic appropriation of specialized labor markets to "create content, solve problems, even do corporate R&D."[45] Howe tied crowdsourcing to Amazon's Mechanical Turk program, a "Web-based marketplace that helps companies find people to perform tasks computers are generally lousy at," such as identifying images, sifting through interpretive information in documents, creating annotations, and more.[46] This is where historical and contemporary treatments of the term "crowdsourcing" diverge: modern crowdsourcing privileges human intervention over computational algorithms, and relies on the dialectic relationship between humans and technology. Although the term "crowdsourcing" was coined for application to corporations, many public participation science research projects have employed elements of crowdsourcing methodologies in order to annotate or collect large amounts of data that could not otherwise be annotated with a smaller number of researchers addressing the problem.

Crowdsourcing is controversial. One of the criticisms levelled at the crowdsourcing method of data collection involves the same criticism of participatory methods: data inaccuracy. Some methods of crowdsourcing are extremely exploitative: at sites such as Amazon's Mechanical Turk, CrowdFlower, Clickworker, or Toluna, for example, workers are often paid well below minimum wage—exploiting labor pools at home and abroad.[47] Data Action does not support using exploitative crowdsourcing.

In many ways the popularity of crowdsourcing was driven by technology that allowed easy access to vast groups of people. The increased availability of smartphones beginning in 2007 with the release of the first iPhone and Android HTC device catalyzed widespread personal, mobile access to the internet. Preceding this was the launch of two major social media networks, Facebook (2004) and Twitter (2006), which would eventually become venues for communicating and sourcing information and data. Yet these sites often only reach part of the crowd because their user base does not represent the wider population. While reaching the entire universe of people is often not the point of data collection projects, getting enough data to provide evidence for a claim is, and crowdsourcing can therefore be an effective technique for collecting data meant to generate civic change. Crowdsourcing projects appear to work best when the groups involved have a stake in collecting the data. Whether it is an interest in helping during a disaster event or concerns for environmental conditions in their neighborhood—personal connections help generate better data.

The Safecast team originally thought to sell the bGeigie fully assembled, but realized that people seemed more invested in the project and contributed data more regularly when they received the kit and had to build it with their own hands. Co-creation of the sensor made them feel like co-owners of the project itself.[48] Perhaps the greatest successes of the Safecast team are the education workshops and programming they perform to teach people to use the devices, but also about the dangers of radiation, how sensors measure, and how to physically make this electronic equipment (figure 2.12). The project was about much more than collecting data, it was also creating a community around building that data. Data Action is not just about data analytics; the process helps build the action needed to create change.

Ultimately, a map of radiation levels throughout all of Japan was created (figure 2.13) that Safecast insists operates outside the politics of nuclear power. They claim that they are neither proponents of nuclear power nor against it, which according to them means the data is more trustworthy as they have no motivation to discredit or exalt the data they generate. There have been questions as to the relative accuracy of the bGeigie devices, but while Safecast acknowledges that the reading may not be precise, they say the purpose of the devices is not precision but a relative understanding of risk, which can help the government take action should there be an environmental threat in the community. Having a completely crowdsourced database for radiation levels in Japan is itself an accomplishment. Gathering complete data in a crowdsourcing project is often difficult, because it is challenging to obtain participants across all geographies or socioeconomic groups.

2.12. Participants of Safecast put the sensors together. Source: Safecast and Marc Rollins, "Safecast Blog Image Compilation," n.d., https://blog.safecast.org/.

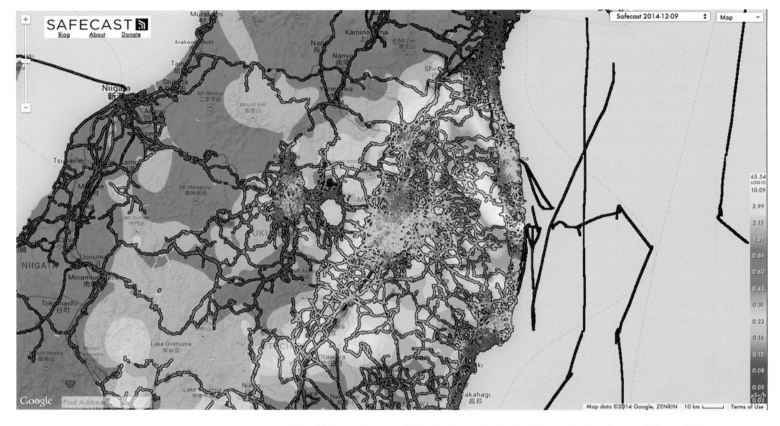

2.13. This map shows radiation in Japan detected by Safecast devices. Source: Safecast, "Safecast Tile Map," 2019, http://safecast.org/tilemap/; Google, "Google Maps Tiles," n.d., via Safecast.org.

Automated sensing projects—those in which machines do most of the work without direct human monitoring—often track subtle differences in the environment. The Rainforest Connection is a standout example: it uses upcycled mobile phones to monitor deforestation rates, eliminating the need for a constant human presence in the rainforest to collect data. The sensors respond to spikes in noise levels that could indicate the presence of chainsaws, bulldozers, or other tools used to cut down trees. In Brazil, where dozens of environmental activists die in a given year defending environmental policies surrounding deforestation,[49] deploying a mobile phone in lieu of a person can save human lives.

Recently many sensing projects have attempted to capture data about informal infrastructure services (such as transit or trash collection) that are not officially provided by local government but are essential to daily life in many rapidly developing cities in the Global South.[50] This is work that has influenced me personally, as I mapped and shared data on Nairobi's informal transit system of minibuses (called *matatus*). The Digital Matatus project, as the work came to be known, had many of the elements important for making data work for policy change. Not only was an original and essential data set developed, it was done in collaboration with the stakeholders and then freely distributed. I use the Digital Matatus project as a case study in chapter 4 to outline the benefits of sharing data, but I want to highlight here the impact that developing the data set had on the community.

The majority of people in the world depend on semiformal transit systems for mobility. From Nairobi to the Philippines, these private systems operate in areas where the government has provided no alternatives, and there is usually no published data on routes or schedules, which if available could be extremely helpful for transportation-planning efforts in these cities. In some cases, a private company may have the data but does not release the information. The Digital Matatus project used cell phones to collect route and schedule data and, with a team of students from the University of Nairobi, was able to generate a comprehensive map of the *matatu* system (figure 4.6). In other words, from a simple tool that almost everyone in Nairobi holds, we were able to generate a database searchable in Google Transit, allowing anyone in Nairobi to map out their transit routes, essentially putting information that now we may take for granted in New York City, into the hands of people in Nairobi for the first time.

The data developed for Nairobi was essential for developing the counter-narrative of the *matatu* system, which was often perceived as a disorganized mess with hundreds of operators. The data showed that the *matatu* owners in fact planned their routes and stops and collaborated among each other to enforce those routes. The data helped multilateral agencies understand that the *matatu* system was well organized, and they could work to develop collaborative projects to ensure safety and improve the quality of transport in Nairobi. In collaboration with the World Bank, the Nairobi city and county government used the data to plan for a Bus Rapid Transit (BRT) system on the most traveled routes.

Data from Citizens: Why Governments Collect It

We most often think of participatory sensing projects as developed outside of formal institutions, but many governments seek to collect highly localized data so they can draft better local legislation. Take, for example, New York City's Department of Health and Mental Hygiene air quality initiative. In 2007, New York had five to ten working air quality sensors throughout the entire city that measured the overall condition of the air but in no way addressed the highly localized conditions that can make air quality worse in one neighborhood as opposed to another.

For many neighborhoods that had highly localized polluting sources, these air quality sensors were just not enough. Questions about air quality were particularly concerning in low-income areas, where asthma-related emergency room visits were higher.[51] So New York City deployed 130 active monitors at different locations throughout the five boroughs. The result created the "the most comprehensive geographic coverage of any urban monitoring network in the US," which is now a model for cities around the world (see figure 2.14).[52]

What is interesting about the project is that sites for the sensors were selected with community input, even though this type of involvement is not common for government data collection projects. Community input is important to using data for action. The Department of Health and Mental Hygiene sought the advice of leaders in each community district; then they created what they termed "systematic site allocation,"[53] which looked at geographic and socioeconomic factors that would make a site more or less suitable for an air quality sensor. Some factors were only considered after

2.14. Results from the studies of New York City air quality translated into policy reform, which in turn produced better air quality, as can be seen on the maps above. Source: New York City Department of Health and Mental Hygiene, "The New York City Community Air Survey: Neighborhood Air Quality 2008–2014," April 2016, https://www1.nyc.gov/assets/doh/downloads/pdf/environmental/comm-air-survey-08-14.pdf.

conversations with community leaders; in fact 80 percent of the sites were determined by statistical analysis, leaving 20 percent to be determined by community leaders. After 2009, fifteen additional monitors were placed that specifically monitor environmental spaces in lower-income, marginalized communities that were previously unsampled.[54] Data analyzed showed that many New York City residents suffered from poor air quality because they lived in neighborhoods where there were high numbers of apartment boilers that used oil known to be a pollutant, and which had not been replaced because of grandfather clauses. The Department of Health had long suspected this was a problem, but the data analysis gave it the evidence it needed to call for a law against these boilers and to create a program that would provide economic assistance to help property owners with their removal.

In low-income neighborhoods like the ones in New York, government sensing projects can build civic trust. When governments design sensing projects with the community, they are arguably much more successful not only because community input can improve the data collection process (as in the selection of sensor sites), but also because the community feels more confident about the accuracy of the data. Perhaps more importantly, public sensing projects that have government backing can ultimately lead to policy change more quickly and directly because of the mutual trust created from working together.

Data Collection during Disasters: Governments and People Depend on It

When disaster strikes, collecting data is essential for humanitarian response. "Humanitarian OpenStreetMap" is an initiative that, in the event of a crisis or disaster, offers free, up-to-date maps locating where problems are occurring and where relief is available or not. Humanitarian OpenStreetMap was employed in the wake of the Ebola outbreak in West Africa in 2013 and 2014 (figure 2.15), enabling activists to map various responses—from the location of and drive time to laboratories, to media coverage of where the outbreak was occurring. The team works with organizations such as the Red Cross, Doctors Without Borders, the United Nations, and other government agencies to develop maps according to specific needs. The 2015 Nepal Earthquake Response Team, for example, aggregated GIS data about Nepal (e.g., areas where aftershocks were occurring, locations of amenities, main roads and buildings) and prepared maps for download and use offline for individuals providing relief on the ground. Many such examples can be found using the OpenStreetMap platform.[55]

In "crisis mapping" citizens actively participate in data collection in the wake of a natural or civic emergency. One of the first tools to address disaster relief through crowdsourcing was Ushahidi, a web and mobile platform designed to provide a way for people to report on what is happening during a crisis. Once installed on a user's cell phone, the Ushahidi platform can be used to send emergency reports from the field via SMS, social media, and web-based reports.[56] The platform will geolocate a position using a cell phone's internal GPS and send the report back to the platform as a point on a map. As more people text in, those who administer the Ushahidi platform might see clusters of reports that indicate the need for water or send an alert about an impassable road.

Ushahidi was developed after the post-election crisis in and around Nairobi in 2007 and 2008, during which the incumbent president Mwai Kibaki was declared the winner of the presidential election but was accused of electoral manipulation by supporters of Kibaki's opponent Raila Odinga. The conflict sparked violence throughout Kenya, much of it in informal settlements where the political rivalry between Kikuyus and Luos exacerbated longstanding tensions between different tribal groups. Many people were trapped in their homes, with no way to obtain food or water, as they waited

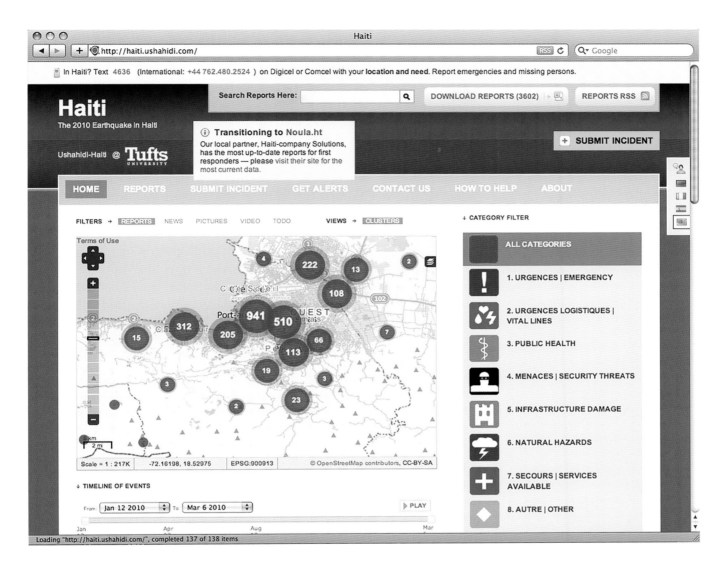

2.15 Screen shot of the Ushahidi website developed for the 2010 Haiti earthquake relief efforts. The map shows the number of reports asking for assistance after the earthquake. Source: http://haiti.ushahidi.com/

for the violence to end. Ushahidi, which means "testimony" or "witness" in Swahili, provided a means by which those in the middle of the conflict as well as remote audiences could collect information about hazards and basic needs, report what they saw via e-mail or SMS, and then have the data mapped in real time and added to an online map on which all issues were logged.

The technology developed for Ushahidi was repurposed for other disasters, including the 2010 earthquake in Haiti, where students at Tufts University obtained the open-source code developed for the Nairobi election crisis and created a platform to be used on the island. At first the site largely used social media to gather people's incident reports. Later with the involvement of Robert Munro, a PhD candidate in linguistics at Stanford University, they created a short code SMS (4636) that allowed people to text in to the platform and let emergency workers know what they needed.[57] Witnesses could text about what they were hearing or seeing, and messages such as "people are trapped in a building on Border and Smith" would appear, which would help focus responders' relief efforts.[58] After twenty-five days, the site had received almost 2,500 crowdsourced complaints, which helped save many lives.

The work done as part of the Haiti earthquake response helped popularize Ushahidi, which was not well known before then. But the relative ease with which the tool was repurposed made Ushahidi integral to what might be called a crisis-mapping movement. In many ways the movement exists thanks to the evangelism of Patrick Meier, who spearheaded the Haiti project at Tufts, went on to launch a crisis-mapping conference, and co-founded the Digital Humanitarian Network. After Haiti, a Ushahidi-inspired site was set-up for the Chilean earthquake in 2010, again by a student group, this one at Columbia University's School of International Policy and Affairs (SIPA).

Ushahidi has been used at the site of many disasters since these early years but has been sometimes criticized for how it can tend to generate expectations that may go unmet. By providing a place to text, it can leave users with the impression a response is on the way. Although Ushahidi did a great job collecting the data, the teams that used the platform did not always have a plan for who would respond to the many queries and expressions of need.[59] How the information would actually generate a response was an afterthought, and the students or volunteers who set up the systems needed to figure out, while still building the technology, how to get stakeholders such as the Red Cross involved. The systems are also criticized for their inability to create a comprehensive picture of a disaster (at the level that would satisfy historians, for example). Only people who knew about the platform would use it, making the resulting data set a bit biased toward those who are more technologically savvy, and leaving the experience of those who did not know about it—or did not have cell service—unrecorded. It is important to interrogate who is missing from your crowdsourced data, as that, too, can provide insights.

One of the biggest problems encountered in crowdsourced data projects is finding enough participants to create a significant amount of meaningful data. When Hurricane Sandy hit the mid-Atlantic region of the United States in 2012, resulting in massive flooding of streets, tunnels, homes, and widespread power outages in- and outside its large population centers, Schuyler Erle, from Humanitarian OpenStreetMap, developed a website tool that would allow volunteers from anywhere to view satellite imagery to evaluate damage.[60] Enlisting an established volunteer group from the crisis mapper network to verify the data, the results of this crowdsourcing project were systematic and, therefore, more accurate than what was seen in previous Ushahidi projects. While the hurricane was happening and during subsequent relief efforts, the Federal Emergency Management Agency (FEMA) leaned on the OpenStreetMap community to help review aerial imagery of the damage to determine the hardest hit locations. Over four thousand online volunteers contributed to this effort, categorizing images of structural damage and helping the agency prioritize relief efforts. Additionally, members of the GIScorps, a group of GIS professionals from all over the country who volunteer during disaster events, helped ground these observations in a more "expert" perspective by evaluating the accuracy of crowdsourced annotations of the aerial images. The application pulled up an aerial image and asked the volunteers to assess whether structures in the

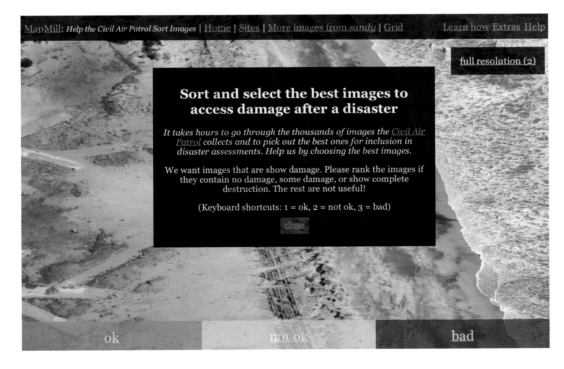

2.16 The interface for Humanitarian OpenStreetMap was site developed to record damage after Hurricane Sandy in New York. Source: MapMill, Image of Map Mill App Courtesy of PublicLab.org, n.d., website, n.d.

2.17 This satellite data shows what participants of the Hurricane Sandy project would use to identify the level of damage in an area. Source: Humanitarian OpenStreet Map Team, https://www. hotosm.org/.

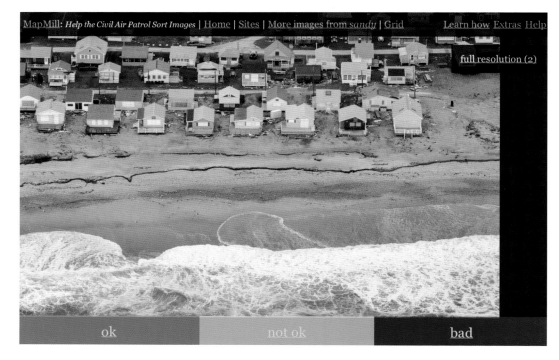

2.18 This map, created by the Humanitarian OpenStreetMap team, shows locations where people used satellite data to map damage after Hurricane Sandy, as well as the extent of the damage (green = good, yellow = average, orange/red = bad).Source: Humanitarian OpenStreet Map Team, https:// www.hotosm.org/.

image were okay, not okay, or bad (see figures 2.16 and 2.17). The results of the combination of crowdsourcing and expertise determined a map of areas to target for recovery (figure 2.18).

Data Collaborations: Achieving Data Literacy and Community

Citizen data collection projects are also educational and improve data literacy, as participants must understand the data they are collecting. Whether the projects involve learning how to build a Geiger counter or interpreting satellite images, all data collection projects include an element of learning. The act of collecting data can be fun, too. Participants get to go out into the field with others who have similar interests and learn how issues affect their community, and they learn the science behind the numbers. In many ways, participatory data collection encourages people to engage with STEM (science, technology, engineering, and math) topics regardless of their formal training in any STEM discipline.

Mobilize, a smartphone-based sensing platform developed in 2010 by UCLA's Center for Embedded Networks and Sensing, is a pedagogical tool for "hands-on, inquiry-based curricular units and teacher professional development for computer science, mathematics, and science high school classes."[61] A 2017 recent NSF grant enabled Mobilize to be integrated into the Los Angeles Unified School (LAUSD) district's curriculum as part of their "Introduction to Data Science" work. For one of their first participatory sensing projects, they developed a tool to install on students' mobile devices, which asked simple survey questions that were largely sociological in nature (e.g., sleep patterns, presence of advertising in neighborhoods, snacking habits). Students then used the data collected to perform statistical analysis.[62] UCLA/ LAUSD believed that this process engages the students more and teaches them not only how to work with data, but also to contextualize the results within their communities. They believe the curriculum "challenges students to develop both computational and statistical thinking skills by learning to ask statistical questions, examine and collect participatory sensing data (and other data collection paradigms), analyze data using R (a statistical program), and interpret and then present their results in written and oral reports."[63]

Participatory sensing projects increase data literacy and teach participants how to gather, clean, analyze, and represent data. Data Canvas: Sense Your City was a data art project run by swissnex San Francisco and Gray Area, an art and tech nonprofit. For six-month period, the Data Canvas team supplied one hundred people across the globe with a DIY environmental sensor kit to assemble and deploy. This was part of an effort to collect real-time environmental data from around the world for a larger data art challenge that invited participants across disciplines to visualize, sonify, and represent the data in different ways. The aim was not to capture accurate information; instead Data Canvas was an exercise in encouraging creativity around what can be done with data. The map that displayed the sensor data was designed specifically for a general public who had little exposure to environmental sensing as a way to encourage people to explore data and feel comfortable approaching and working with it.

Citizen Data Collection Projects Do More Than Collect Data

The design of data collection itself should be collaborative because including diverse actors during project planning can help create a community around a specific topic—even if all the stakeholders don't participate in the data collection itself. The Safecast project, for example, gathered people interested in radiation, and a support network was grown through teaching participants how to make the sensors that would measure radiation. OpenStreetMap generated a community of open-source mappers dedicated to making data free and devoted to crowdsourcing to perform that work. Community building through data collection can even happen when governments implement the project. This was true of New York City's Department of Health and Mental Hygiene's air-quality sensing project. The project was designed to obtain feedback from the community about sensor locations; working with the community in this fashion helped residents feel integral to the project's success and helped build trust in the outputs. This type of community building was also witnessed in the Public Lab project in Louisiana, which rallied people around developing a counter-narrative to BP's reports about the oil spill.

By building participatory sensing projects we build communities, interest groups, and counter-narratives essential for civic change. In the Beijing Air Tracks project and the Pennsylvania fracking project, the data provided evidence of what almost everyone knew to be true: air quality in China is horrendous and fracking causes environmental externalities. In the case of Beijing, the ensuing debate transpired in newspapers across the globe. Although sparking a global debate is not always possible for citizen data collection, the fracking project shows that a local debate can still be effective; Pennsylvania now has better enforcement of air quality measures.

Collecting data also leads participants to truly ask and understand what the data means. What is air quality? What is particulate matter? What units do we measure radiation in? In Beijing the AP reporters became air-quality data experts, learning what micrograms per cubic meter means and that levels above 50 do not meet WHO standards for developing countries. They learned that real-time sensing devices measure PM thanks to light sensors that measure the shadow of the particles. Learning can even make the project more engaging, so much so that the designers of some environmental sensing projects ask participants to assemble the sensors themselves—as the project that mapped radiation in Japan illustrated. OpenStreetMap educated a generation of novice mappers about how to develop data with GPS units or trace data from satellite images. During the Safecast project measuring radiation in Japan, the organizers found that when participants made the sensors, they were better at maintaining both the data and the sensors' functionality.

Building data is one of the essential ways we create evidence in planning projects, particularly wherever the data is tightly controlled by governments or does not exist. Yet in a world where almost everything is being *sensed* in some fashion, we can also obtain data for policy action by collecting it from unusual sources found on the web—from social media to restaurant reviews. Roughly 80 percent of the data now stored globally is privately owned, but this data is sometimes easily acquired by scraping it from websites. In the next chapter we will look at how collecting this data and repurposing it for urban analytics can generate policy change.

HACK IT! USING DATA CREATIVELY

The term *hacker* conjures up images that range from nefarious Russian spies trying to influence the United States elections to computer geeks sitting in their dorm rooms eating junk food and playing *Minecraft*. For many, hacking means following the hacker ethos, summarized in the preface of *Hackers: Heroes of the Computer Revolution*, by Steven Levy: "sharing, openness, decentralization, and getting your hands on machines at any cost to improve the machines, and to improve the world."[1] Yet the Hacker Code of Ethics, as first documented in Levy's book, leaves ample room for interpretation. This means "hacks" can be used toward the development of both "good" and "bad" applications. Perhaps more importantly, how "hacks" are applied represents the values of those that develop them.

The hacker culture, with its earliest roots at MIT in the 1960s, had a huge influence on my graduate and professional life as well. For me, and for many MIT alumni, hacking represents a lifestyle built on the intellectual challenges of creatively overcoming limitations, often by using technology. But it can also simply refer to innovations of all sorts (e.g., "life hacking"). For example, there is a whole website dedicated to creative use, or hacking, of IKEA furniture; Ikeahackers.net has instructions to make your bike trailer look like a coffee table and how to hack a portable shower tent for your RV. I like to think of hacking as a positive and empowering way to tackle the world's problems using ingenious innovation: for me it's about being creative in the way data is accessed and used, not about stealing data or capturing information through criminal means. The phrase "hack it" can be applied to many arenas—political, progressive, criminal, and those in between. I don't intend this chapter to be a critique of or debate about hacktivism, but rather a call to action, following the most idealistic interpretation of the Hacker Code of Ethics.

Today there is more data in the world than ever before. How that data is used—whether to sell products or to help answer questions important to society—has become a much-discussed topic, one that often begs ethical consideration. When it comes to using big data for policy change, numerous research articles delve into all types of interesting applications, but many miss important methodological elements. Namely, to be able to use big data for the improvement of society, we must work with policy experts to ask the right questions, verify our results, and interrogate biases in the data itself. Working with data in this way not only provides better results, but it's also ethical. Data represents people, and making sure those people are accurately represented helps to reduce the ways that data analysis can marginalize populations.

This chapter provides some examples and some cautionary tales that concern the use of data openly available on the web, and how that data can be applied for policy change. Sometimes, data found on the web is the only data that exists on a topic, and using this data to answer questions important to society means that we have to interrogate its biases as well as potential ethical concerns. Some criticism has surfaced about the messiness and bias of data scraped from websites like Twitter. But all data is biased, and that doesn't preclude us from using it. If we understand this bias, we can better map the ways it can be analyzed ethically; in fact, investigating biases in data often provides new insights. That's why identifying people and places missing from data sets can not only expose how we often systematically forget certain groups, but also help us apply new policies to address their needs. This chapter asks big data enthusiasts to work with experts, to shape questions that will help society understand underlying social-spatial patterns, especially patterns that to this point have been neglected or unseen. Instead of using big data to sell us something, this chapter asks us to apply it to improve society.

The Ethics of Using Data We Find on the Web

Among the most important ethical concerns about the use of big data today is one that involves the ubiquitous "terms of service" that companies all over the world ask us to "sign" before to using their online services: as a result we provide our consent for these companies to use the data we contribute to their sites; but that doesn't mean we approve of some of the ways the data is eventually used.

Data sets amassed by private companies include vast amounts of detailed information about our personal habits and behaviors that can easily be turned against us. The Facebook/Cambridge Analytica scandal provides a smoking gun for how the data we contribute online is used without our explicit consent to sell us political agendas and offer fake news. Here's a brief recap of how the situation played out.

Cambridge Analytica, a now-defunct British consulting firm, developed political ads for the Kenyan leader Uhuru Kenyatta, Donald Trump, and proponents of Brexit during its initial referendum, among others. The firm used personal information from close to 87 million Facebook profiles, which it acquired without account-holders' explicit permission, to generate psychological profiles. These profiles were then used to target political advertisements that drastically changed the political landscape of Kenya, the United States, Great Britain, and perhaps other countries still to be revealed.[2] Questions still exist about whether Facebook knew that the data was collected against their terms of service and to what extent, but it has been confirmed that data was acquired through an application developed by Aleksandr Kogan, a lecturer at the University of Cambridge's psychology department. Kogan used an app called "thisisyourdigitallife" to capture data on close to 275,000 Facebook users who consented to participate in a study but didn't know that the application was also capturing information about their friends. This allowed the firm that Kogan owned, Global Science Research (GSR) to amass data that Cambridge Analytica later purchased.[3] While reporting on the unusual Brexit results, Carole Callwalladr stumbled upon the details that uncovered the Cambridge Analytica scandal, one of the biggest data scandals in history.[4]

Facebook pushed much of the blame for the misuse of data onto Kogan, saying he violated the terms of service without its knowledge. Kogan believes Facebook made him the fall guy and has started a defamation case against Facebook for its campaign against him during the scandal.[5] Curiously, many of the characters in the scandal had previous relationships with each other long before the data was amassed, which has raised further questions during US congressional and UK parliamentary hearings about the complicity of the parties. Did Facebook hire Kogan to capture user data for use outside the company's data-use restrictions? Did Facebook help start Cambridge Analytica? The answers to these questions have not yet fully come to light. Perhaps all groups involved are to blame. It is certainly problematic that a university researcher, who ought to have upheld the ethics of research set up by the Institutional Research Board (IRB), was responsible for the way in which this data was amassed. Although he was performing this research through his private company, not through his university, the assumption of propriety that comes from being affiliated with an academic institution may have influenced participants to join the study in the first place. At the same time Facebook should have protected the data better, it should not have been possible to amass Facebook users' data to be used in this way. Cambridge Analytica, for its part, violated a different series of ethical concerns, including the use psychological targeting to deliver news from questionable sources to sway public opinion.

As the scandal was unfolding, I understood that much of what I was writing in this chapter about the creative use of data might sound cavalier. But my intent was quite the opposite. I am interested in how we can recapture the data we collectively contribute to private companies and use it for the betterment of society. That question, however, sets up a series of other questions. For instance: What exactly does the betterment of society mean and for whom? Is it ethical to use data we find on the web? So let's take a moment to examine how we deal with the numerous ethical issues associated with data that is used to address issues important to society.

"Ten Simple Rules for Responsible Big Data Research," an article written in 2017 by leading scholars in the field of big data and ethics, foreshadows some of the serious ethical concerns that came out of the Cambridge Analytica case: its authors acknowledge that big data can be used to help society as long as certain boundaries are acknowledged—and the they provide guidelines (in the form of a 10-point checklist) for doing that.[6] I implore all of those who use data (big and small) to abide by these tenets. They appear in table 3.1 alongside my interpretations of the authors' intent.

One tenet that resonates strongly for me is the idea that "data are people." I have always believed this to be true. I think that to be ethical with data it is essential that we show the results of our data analytics to people *in* the data to ensure our analysis rings true to them. Perhaps more important is that the work we do with the data is not used to marginalize groups of people; we need to understand how our interpretations might have missed the mark. It

Table 3.1

	Proposition	Interpretation
1	Acknowledge that data are people and can do harm.	Here the authors remind as that most data is about (or will affect) people, even if people are not captured in the data itself. As you perform research it is important to think about the effects your analysis may have on those people.
2	Recognize that privacy is more than a binary value.	The perceived public-ness of data doesn't mean when analyzed it might not expose personal information. Sometimes data when analyzed collectively can show trends that infringe on our privacy. An example used is historic redlining practices described in the first chapter of this book.
3	Guard against the possible re-identification of your data.	The authors remind us of numerous studies in which the researchers thought data was anonymized sufficiently but was not, and they recommend some guidelines for releasing data.
4	Practice sharing data ethically.	Here, the authors remind us that the terms of service provided by big data companies, which include a range of data from your cell phone traces to your credit card information, are broad and not necessarily ethical. Even if we have the legal right to use data, it does not make the use ethical.
5	Consider the strengths and limitations of your data; big does not automatically mean better.	This reminds us that data is full of bias and that it is important to look at data lineage and provenance to ensure it is suitable for the questions you seek to ask. Is there missing data?
6	Debate the tough, ethical choices.	Institutional Research Boards (IRB) are not the authority on the ethical use of data. Although they help by providing guidelines, the question of what data uses are ethical leaves a lot of room for debate.
7	Develop a code of conduct for your organization, research community, or industry.	This helps insert and maintain ethical thinking and behavior directly into the workflow to remind everyone of its importance.
8	Design your data and systems for auditability.	The ability to audit data use makes it possible to trace the accuracy of reports based on that data.
9	Engage with the broader consequences of data and analysis practices.	Remember that big data research can have broader implications for society, from the amount of energy used to store the data to its ability to expose inequities.
10	Know when to break these rules.	Here the authors acknowledge that there are times when breaking the rules means saving a person's life or exposing social harm, and that sometimes a checklist needs to be thrown out the window for the greater good of society.

Source: Developed by author based on Matthew Zook et al., "Ten Simple Rules for Responsible Big Data Research," *PLOS Computational Biology* 13, no. 3 (March 30, 2017): e1005399.

is through this conversation that I believe we can create community through data analytics, but more importantly help to avoid some of the numerous ways data can unintentionally marginalize populations by marking them as "other" through data analytics.[7]

The Cambridge Analytica scandal is not the first time that we have found our privacy exposed at the hands of large private companies. Just this morning, as I prepared to start writing, I read an article with the following headline: "*Google still keeps a list of everything you ever bought using Gmail, even if you delete your emails.*"[8] The author, Todd Haselton, goes on to detail how the company does indeed keep a record of everything you've purchased, ostensibly to help you track a package or reorder an item using Google Assistant. But no matter what you do you can't delete these data traces, even though Google claims you can remove them by deleting emails that document the transactions. Haselton tried getting rid of a decades-worth of sent and received email in his own Google account, but the traces of purchases made years ago still existed on the Google Purchases page. This example shows we often have little or no agency over how out data is being used. What's perhaps more troubling is that Google is legally permitted do this, whether it seems ethical or not, since we sign away our rights to our information when we click to accept Google's terms of service. Whether it is indeed ethical for Google to behave this way is something that society needs to determine over time, and we must work with our governments to generate laws and regulations around the ethical uses of data. In the meantime, while those laws are developed, we all must be aware of the ethical considerations of using data and take up a Hippocratic oath to *do no harm* in the work we do with data.

Where Do Researchers Get Big Data?

Big data can be obtained in many ways; for example, by scraping it off a website and writing queries to an organization's Application Programming Interface (API), which allows users to access parts of its internal database. It can also be bought from companies, such as LiveRamp, a company that collects our GPS traces from cell phone apps.[9] In this chapter, I am asking all readers to think about the ways we can use this data ethically and responsibly to answer questions that can help society.

Scraping data refers to the automated process of collecting data available through web portals. Even the most novice programmer can develop database queries to generate a complete data set from API queries. Companies often limit how much data can be extracted through their APIs. Twitter, for one, only makes 10 percent of its data available through its API. If you want the full data set, often referred to as the firehose, you need to apply through Twitter's research program. Only a few academic groups across the country have access to Twitter's data. Companies make their data available through their APIs because it's good for business. For example, the reason I can see Yelp recommendations on my Google Maps page is because of connections made through their APIs—This benefits both businesses as it brings more customers to their respective sites. Companies allow researchers to use their API for similar reasons: the innovations help advertise their products and show a value they might not have even realized in their data sets.

One of the early examples of using social media sites for research was a project I developed in 2011 involving Foursquare, a location technology platform.[10] I re-created the Foursquare points of interest (POI) database for eight cities by sending over 250,000 API requests to find Foursquare venues in each city. I emailed Foursquare to let them know what I was doing; their response was "have fun." As long as I wasn't reselling the data, they were happy to have me do the work. Also, it would save them time—they didn't want to donate the staff time needed to give us the data. The project turned out to be good press for Foursquare; the platform was recognized not only for helping us understand land-use patterns in Mexico and Brazil, data that was difficult to acquire in 2011, but also for providing their data for a public good.

Governments also make data available through their APIs, and sometimes that is the only way to get that data. In 2017 one of my colleagues wanted data about eviction cases in Boston. He sought the data via a Freedom of Information Act (FOIA) request from the Boston Housing Court but was told it could take months to get. He couldn't understand why it would take so long; cases were recorded and entered in a database that was accessible online. So he decided to write code that would query the database one record at a time and read the results to a text file. He ran his code overnight and ultimately created a complete copy of the database. Court staffers were grateful when he told them what he'd done. They, too, had needed access to that database—in fact, they'd been unable to find out who was in charge of the online housing court data, and therefore unsure where to send my colleague's data request.

It's important to note that in both the Foursquare and Boston Housing Court cases the researchers contacted the data owners as soon they'd collected the data to let them know how it was collected and how they planned to use it. This follows the "Ten Simple Rules for Big Data Research" I mention earlier in the chapter. It's essential to let the original data owners know how you intend to use the data for numerous reasons. First you want to determine the accuracy of your data as well as its lineage: How complete is it? Does it represent all the data or just parts of it? You also want to make sure that the way you use the data fits within the organization's ethical practices. You don't want to surprise them—nobody likes to find out from someone else that their data was used in ways that could be controversial. Finally, the data owners may be able to help improve your research. Whenever I scrape data from the web for a research project, I make sure to inform the organization I got it from. I might not always hear back, but at the very least it gives them an opportunity to respond to how the data is being used to make sure I am using it ethically.

In many contexts, social media sites are the sole source of openly available data about *points of interest* (POIs). This is especially true in cases when the government decides whom they will or won't share data with—after all, data can serve to analyze issues important to society. In China, for instance, the government controls data so tightly that even local planners don't have access to it. My team working on the Civic Data Design Lab's Ghost Cities project made creative use of social media APIs in China to obtain data needed to evaluate the extent of housing vacancies in China and estimate a potential economic risk. Efforts like this are crucial because it is difficult (indeed nearly impossible) to acquire data about many aspects of the Chinese economy—especially the extent to which data on housing vacancies can indicate an economic slowdown. Understanding risks associated with the housing market, which has been an important economic driver since China opened its economy in the late 1980s and early 1990s, can help those inside and outside China understand their economic future.

Applying Unusual Data to Find Housing Vacancy in China

Data analysts should have the ability to use data collected by private companies for a public good—especially when it can lift up the voices of those on the margins of society. In this chapter, I use my research lab's Ghost Cities project to illustrate using *data for action*—in this case, taking data built by scraping a social media site to estimate housing vacancy. More importantly, our work on China's ghost cities involves people. It shows a process of data analytics that asked those represented in the data to evaluate the results and provide potential solutions to address the high rates of vacancy found in their communities. By involving local Chinese planners in our data analysis, we hoped that our work would be ethical, but perhaps more importantly we anticipated making an impact beyond the walls of academia.

To understand this project, let's first look at how China uses the housing market as a measure of economic growth and at the same time encourages its citizens to use housing as an instrument of personal, financial investment. In 1991, the Chinese government created the Housing Provident Fund, in which employers matched mandatory employee investments (sometimes up to 5 percent) in ways similar to how many 401Ks work in the United States.[11] The matching funds can only be used to purchase a house; many attribute the nearly 90 percent homeownership rates in China (compared to 65 percent in the US) to this policy. Chinese people, said to be "good savers," use their money to invest in second and third homes that can provide returns of up to 10 percent in places like Shanghai, as compared to investments in the Shanghai Stock Exchange, which lay flat with no gains from 2001 to 2010.[12] Many of these second and third homes lie vacant, however, and they continue to increase in value even though no one lives in them.

High vacancy rates have become a serious concern in many Chinese cities, especially in those experiencing an economic slowdown, which some say started as early as 2008 during the global economic crisis.[13] The phenomenon, often called "ghost cities," usually describes whole cities of vacant properties; but it can also mean pockets of existing second-and third-tier cities with high vacancies as a result of poor planning that results in lack of access to jobs or amenities.[14]

Measuring the extent of these vacant areas has been difficult.[15] Although it is believed that top levels of government have a better understanding of the problem, local planners, real-estate developers, and researchers have limited information on how to view, let alone address, the phenomenon. Scholars have attempted to measure these vacancies through various means, including assessing light at night and tracking internet use, but their studies have been hampered by imprecise or unavailable data.[16]

The Civic Data Design Lab (CDDL) set out to test whether it was possible to measure vacancy by using data downloaded from Chinese APIs. Over the course of three months in the fall of 2016 the team developed code to download data from several of China's top social media sites: Dianping, Amap, Fang, and Baidu. The process took some creative coding, since APIs in general only allow access to one record at a time. Our research team contacted each social media site we wanted to acquire data from to ask if they would provide us with the data. The only organization to respond was Baidu; after an initial attempt to share the data with us so we wouldn't have to access it through their API, no one in the organization followed through or returned our emails, and we ultimately lost contact.

So, in order to get the data, we sent thousands of latitude-and-longitude ("lat/long") points (roughly fifty meters apart) and invoked multiple API user keys on a timed-request interval to avoid reaching our API calling limits (figures 3.2a and 3.2b). The APIs we worked with had a function to return POIs near a given lat/long point (figure 3.1 shows a hypothetical "fish net" grid of points). For example, in the case of Baidu (China's predominant search engine, equivalent to Google), a request to the API involved a line of code that included the lat/long and a specification of the type of interest point (residential location, restaurant, etc.). The API returned a string of data that included the type of POI, its name, lat/long, and street address. Each time one lat/long query was sent, the data returned from the API was added as a row in a local database table. Once the information was successfully downloaded for our test city of Chengdu, we captured data for six more second- and third-tier cities: Shenyang, Changchun, Hangzhou, Tianjin, Wuhan, and Xi'an.

After extracting the data our team began to develop a predictive model to identify pockets of vacancy in existing cities.[17] After a few failed attempts, including one process that looked at activity levels based on Weibo data (China's Twitter equivalent), the team homed in on a model shaped by the idea that healthy communities have good access to amenities—restaurants, banks, schools, malls, beauty salons, grocery stores, and other commercial establishments.[18] According to quality-of-life researchers, amenities help determine housing decisions.[19] Edward L. Glaeser, Jed Kolko, and Albert Saiz (2001), for example, argue that the future of cities will depend on the availability of such amenities and the consumption capital they generate.[20] Similar work in China has shown that there is a positive relationship between housing prices and urban public amenities such as public transit, parks, and educational institutions.[21] Given how the presence of amenities suggests the health of a community, we assumed that residential areas with no amenities in close proximity were likely to be vacant or partially occupied.

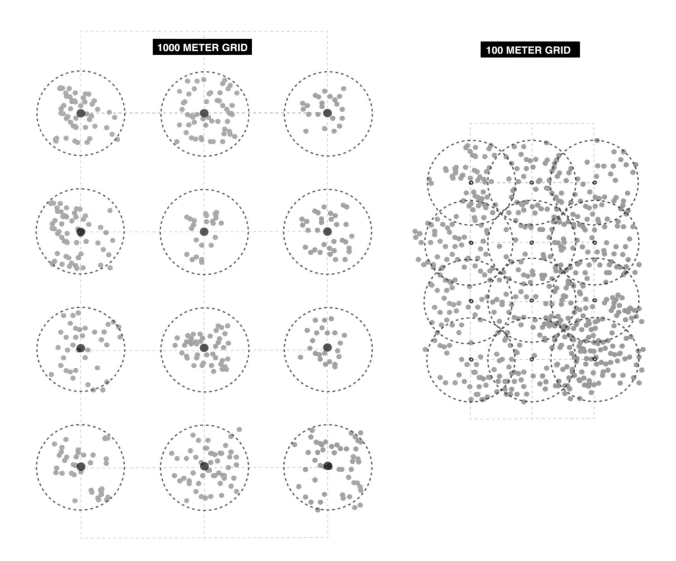

1000 METER GRID

100 METER GRID

Next, from the mounds of data we had already extracted, we needed to identify the right data set to support a model based on access to amenities. The Dianping (Chinese Yelp) data listed more amenities than Baidu's map application, even the very small local stores, so we used Dianping's data to identify the geographic location of the amenities.[22] Our analysis also showed that residential locations scraped from Amap (China's MapQuest) were often more accurate than those scraped from Baidu, so we combined these two data sets to identify the location of residential buildings. The data was cleaned to ensure there were no duplicates. Determining the accuracy of data scraped online is essential for understanding its potential use.

3.1 Determining the spacing between latitude and longitude points is important when making calls to APIs. This diagram shows that if the spacing is too far apart, you will not get all the locations in the database; the API will stop searching after it returns roughly fifty results. There must be overlap in the results returned, shown in the diagram on the right. This can be done by moving the latitude and longitude points closer together. Any duplicate results can easily be deleted. The spacing is dependent on the density of the area you are working. Spacing needs to be tested for all new projects. Source: Sarah Williams

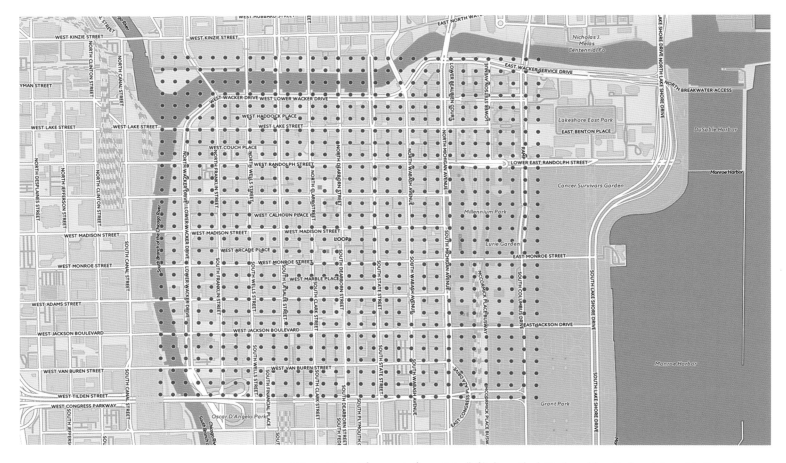

3.2a and 3.2b The map in figure 3.2a (left) shows the latitude and longitude points sent to the API; the map in figure 3.2b (right) shows the data returned by the API. Source: Sarah Williams

© OpenStreetMap contributors © CartoDB,CartoDB attribution

We drew from Hansen's gravitational model of spatial accessibility (figure 3.3), the foundation of many different accessibility measures concerning transportation costs such as access to health care,[23] employment and job opportunities,[24] groceries,[25] and shopping.[26] The model allowed us to develop a measure of accessibility given a supply and demand for amenities.

CDDL modified this model for the Ghost Cities project to look at access to amenities, and then we assigned each residential location an amenity score. To calculate the amenity score, we divided each city we studied into small neighborhoods (300 m²), roughly the size of a typical, gated community in China. We used the geographic center of each neighborhood to measure the

$$A_i = \sum_j \frac{S_j}{d_{i,j}^{\beta}} \qquad \text{(eq. 1)}$$

Where β = exponent describing the effect of travel times between zones

$d_{i,j}$ = size of the travel between zones i and j

S_j = The size of the supply or capacity in zone j

In this model, β is often referred to the 'friction' coefficient, representing the difficulty of travel between zones i and j, whereas S represents the capacity for each amenity. As A_i above is only a 'supply' index, other accessibility measures include a demand component written as:

$$A_i = \sum_j \frac{S_j}{d_{i,j}^{\beta} V_j} \qquad ()$$

Where

$$V_j = \sum_k \frac{P_k}{d_{j,k}^{\beta}} \qquad ()$$

V_j = The demand for amenity j

β = exponent describing the effect of travel times between zones

P_k = Population in zone k that would be using the amenity k

$d_{j,k}$ = size of the travel between zones j and k

3.3 These equations, based on a gravitational model of spatial accessibility, represent the model CDDL developed for to create an amenities score. Source: Sarah Williams et al., "Ghost Cities of China: Identifying Urban Vacancy through Social Media Data," Cities 94 (2019): 275–285.

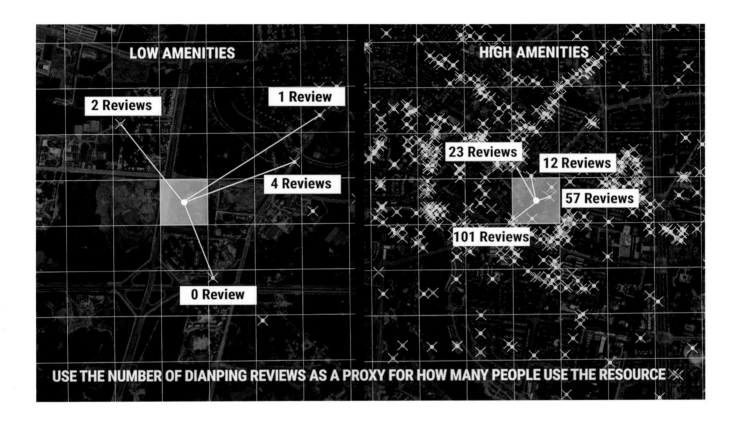

LOW AMENITIES | HIGH AMENITIES

2 Reviews 1 Review 4 Reviews 0 Review

23 Reviews 12 Reviews 57 Reviews 101 Reviews

USE THE NUMBER OF DIANPING REVIEWS AS A PROXY FOR HOW MANY PEOPLE USE THE RESOURCE

distance to each amenity (figure 3.4). Our modified model took into account that folks living in suburbs and rural areas might need (or be willing) to travel farther to a restaurant or school.[27] The model also used Dianping's ratings to identify the popularity of an amenity. Locations with no ratings in Dianping were considered underused, and their amenities score decreased. Once the data was gathered for each neighborhood grid, that grid cell received an amenity score. We removed residential locations that are clearly thriving. Then we removed cells with amenity scores above the mean (see figure 3.5), and were left with potentially vacant lots. After further refinements, we obtained low-amenity-score clusters identifying underused land or ghost cities (figure 3.6).

3.4 The Ghost Cities model starts by measuring the distance to all amenities. Here it is measuring the distance to restaurants, taking into account that those living in the suburbs might be more willing to travel farther to a restaurant. *Source*: Williams et al., "Ghost Cities of China."

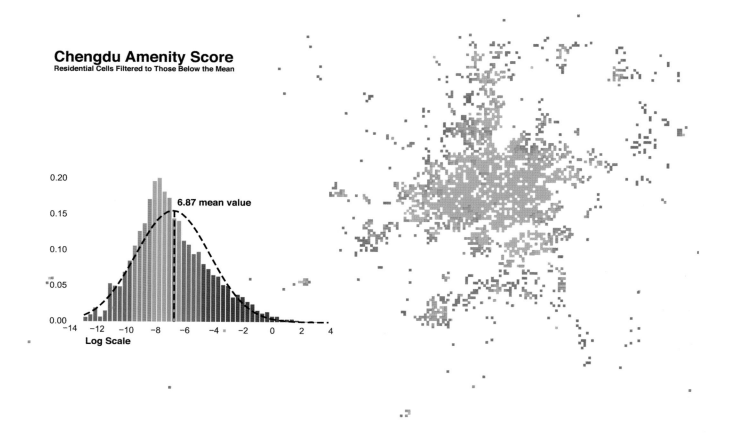

Chengdu Amenity Score
Residential Cells Filtered to Those Below the Mean

6.87 mean value

Log Scale

3.5 A map of the results for the Ghost Cities model in Chengdu after we filtered for those grid cells with amenity scores below the mean. Source: Williams et al., "Ghost Cities of China."

The Data Action method asks data analysts to check the accuracy of their analysis with qualitative data. In the Ghost Cities project we used drones (figure 3.7), surveys, and photographs on visits to Chengdu, Tianjin, and Shenyang. The team also interviewed local people in those cities (figure 3.8). These site visits proved that our model identified all types of underutilized land, which we defined as residential locations that were not meeting development potential, either because the land was unoccupied or construction was stalled. It was important to shed light on these areas since they clearly needed intervention from local actors, including government, to make them viable places to live.[28]

Ghost Cities of China

Chengdu

Ghost Developments
Amenities

The amenity score

Low   Mean(-6.87)   High

$$A = \sum_{k=1}^{8} \sum_{j=1}^{10} \frac{w_k * R_{j,k} + 1}{D_{i,j,k}^2}$$

where
w_k = weights assigned to each amenity category k for 'commercial' vs 'institutional'
$D_{i,j,k}$ = Euclidean distance between each grid i and jth amenity i
$R_{j,k}$ = Number of Dianping ratings for the jth amenity in category k

3.6 Results of the Ghost Cities model in Chengdu, China, show red cells
that mark residential areas we identified as underdeveloped (or ghost) cities.
Source: Williams et al., "Ghost Cities of China."

3.7 The project team, shown here working outside of Chengdu, used drones to ground-truth the grid cells that were marked as ghost cities. Source: Civic Data Design Lab, MIT.

A large portion of our results identified residential locations that may have been developed five years earlier. The government informed us that they believed people would eventually move to these sites, but that it is unlikely to happen without access to schools and other amenities. In time, they say, the amenities will come. We further determined a number of locations where ground had been broken but construction had stalled; semi-vacant housing often surrounded these undeveloped sites. We found vacated properties, remnants of the various building styles indicative of China's communist past; for instance, tower complexes lay vacant and with no one living there, and the amenities that once supported them no longer exist. Finally, our team identified one entire satellite city outside of Shenyang that lay largely vacant (figure 3.8).

Our site visits verified that the model did, in fact, identify a type of vacancy, but we learned much more about possible solutions to vacancy

3.8 The photos here are a sampling of the vacant and underutilized developments found by using the CDDL's Ghost Cities model. Source: Civic Data Design Lab, MIT.

problems by identifying the missing amenities that might have contributed to the vacancy in the first place. Let's say that a residential location received a low score for amenities because the nearest school was far away; knowing this, the urban planners in the city could decide to build a new school.[29]

Meeting with local scholars and planners is integral to the Data Action method. In China, such meetings (figure 3.9) were an essential part of verifying our model for two important reasons. First, they helped the CDDL team confirm that the model represented what local researchers understood about the phenomenon. Second, our team wanted to learn whether having such results would be helpful for those engaged in planning practices in China. We contributed to the local administration by developing an interactive website (figure 3.10) to visualize and communicate the results on a map of each of the seven cities. Exploring the website not only allowed the results to emerge, but it also helped teach those who interacted with it how the quantitative model generated the amenity score for each residential cell (figure 3.11). As expected, in thriving areas of the city, the graph shows that people have quick and easy access to amenities, whereas in the red squares (or ghost cities) many amenities, including school and groceries, are prohibitively far away. The website allows anyone who interacts with it, including urban planners and real estate developers, to determine which amenities are lacking in each residential location. This data-interactive site creates a mechanism by which to share the data and model, making it easier to understand and explore the insights it offers. The website helped facilitate our conversations with the local planners, providing context for the reasons behind the phenomena our results highlighted.

One element of the Data Action principles is that models are not complete until reviewed by the people described in them. In the Ghost Cities project, CDDL researchers shared visualizations of the results with local stakeholders, urban planners, real estate developers, and other researchers, asking for their impressions and feedback. Yixue Jiao, the senior urban planner at the China Academy of Urban Planning and Design, said: "As one might expect, vacancy is controversial; while many planners know [ghost cities] exist, they come from directives from higher levels of the government," making it hard for local planners to push back their development.[30] Although aware of the situation, he felt there was little he could do to change the government's course. Some urban planners we interviewed discussed how the decisions for where to build seldom make use of big data to analyze appropriate locations; they use

Hongyu LIU

Professor of Real Estate and Construction,
Director of Institute of Real Estate Studies /
Tsinghua University

Vice-general Manager of Sales /
Baoneng Real Estate Development Co., Ltd

Qidong LI

Senior Urban Planner /
China Academy of Urban Planning & Design

Yixue JIAO

Secretary General /
Shenyang Real Estate Development Association

Xijing QI

Dean of Business School /
Shenyang Jianzhu University

Yachen LIU

Xiaoke ZHANG

Deputy Director of Yuhong District /
Shenyang Planning Bureau

3.9 The CDDL team took these photos of academics, planners, and real estate developers after showing them the results from the seven ghost cities included on the interactive website. Sharing the website information made it much easier to discuss the model and provided candid conversations about the results of the model. Source: Civic Data Design Lab, MIT.

3.10 This screenshot from the ghost cities interactive website (http://ghostcities.mit.edu) shows the data that went into the model. A highlighted red ghost cities square above an area indicates a low amenity score because it is far from a KTV (karaoke club), school, medical facility, or grocery store. This data can also be read on the interactive graph on the right-hand side of the screen: it includes a street view of the location of the red square. Source: Civic Data Design Lab, MIT.

"theories" or "book knowledge" of what the market might sustain, rather than what is currently observed on the ground. While the planners recognized the accuracy of our results, they felt a limited ability to change the course, even with the data analytics that were presented: many urban planning decisions in China are made by the central government and filtered down for local organizations to enact. It is often a one-way conversation.

The maps fascinated the Chinese real estate developers. A few developers explained how the vacancies we exposed also illustrate China's real estate bubble: if it burst, they believe, citizens encouraged to carry multiple

3.11 A user interacting with the website the Civic Data Design Lab created to present results of the Ghost Cities project. Source: Image by Chaewon Ahn

mortgages would feel the greatest impact. Chinese investors don't always understand their mortgage risks, these developers explain, because banks are officially tied to the government in China, and there are reasons for them to insulate investors from potential risk. In other words, the government will certainly bail out banks for bad loans.

Developers acknowledged that the maps illustrate the mismatch between supply and demand in the Chinese housing market, but explained that the economics of the housing market in China works differently than it does in other places in the world. For the Chinese government, the economic benefits, such as job creation associated with building the development, outweigh the five- to ten-year lag in people purchasing the homes. Researchers believed that addressing oversupply was essential to ensuring a healthy economy and healthy cities: having many cities remain partially vacant increases the difficulty of attracting residents without major government interventions. Local urban planners mentioned that these interventions should not be hard to accomplish; the government simply can tell a company to move to a specific location or develop a school, actions that would make these viable places to live in. Yet the higher levels of the Chinese government have not enacted this solution. In sum, conversations on the ground showed that while the model successfully determined vacancy, the next step would be learning how to create action from the results. Unfortunately local Chinese planning officials are subject to the directives of higher levels of government, which makes changing how development occurs politically complex.

Perhaps more important than identifying vacancies, the model developed for this project shows how scraping data from social media can provide a picture of the current state of development, one not previously available even to local Chinese planners. The Ghost Cities maps that CDDL developed for this project identify risk in the Chinese real estate market. We believe this is akin to having maps that identified the foreclosure crisis in the United States before it even occurred. The CDDL team hopes local Chinese planners and real estate developers will use the interactive web tool as a guide to address vacancy in some of the areas identified.

Finding the Right Open Data

It is important to note that the Chinese government could certainly estimate the extent of vacancy from data only it has access to (such as electricity records), but it does not share this information with local-level officials who could use the data to make decisions about local development. The government controls the release of data because it wants to steer the course of development. Even though there has been a move toward a more market-based real estate development economy in China, from the perspective of the Chinese government, sharing data with real estate developers could have unknown outcomes and, ultimately, the possibility that the Chinese government could lose control over housing development, a risk they are unwilling to take. Real estate developers therefore have to rely on the government to tell them where to build and the risk for developers whose projects don't work well is minimal because the government has an incentive to bail them out. This is because China's banking industry is closely intertwined with its government. Therefore, the government does not want developers to default on their loans because it would cause instability in China's economy.

Determining the data that would best expose vacancy in China took some time because we needed to investigate the potential biases and ethical concerns of the data available to us. Early in our project we hoped that measuring human activity levels using the Chinese version of Twitter (Weibo) would provide conclusive evidence that an absence of human activity indicates the presence of ghost cities. But after collecting all the data accessible online, we realized it was too messy to use. So we dug deeper, only to discover that not everyone uses Weibo: those people were not showing up on the map. To further complicate the problem, only about 1 percent of Weibo posts were geotagged, which severely limited the data set. It turns out that government control over social media companies in China results in data that is highly censored. CDDL's work is all the more challenging because the government deletes Weibo information, which skews the social reality that the data represents.

Indeed, the CDDL had first become aware of these issues when working on a project to expose citizens' impressions of air pollution in China. For this project we used the online version of Weibo to find out what people say about pollution. We found almost no results, which was surprising because at that moment the news was flooded with discussions about Beijing's air quality. When discussions of pollution did surface on the Weibo home page, they were often deleted, almost instantly or within hours. This type of censorship makes it difficult to use social media data to make sense of controversial topics. CDDL's experience underlines how important it is to determine the accuracy of data gathered online, to find ways to identify gaps in the data, and account for missing data when drawing conclusions.

Embracing Biases in Data to Tell New Stories

Interrogating the ways in which a data might exhibit bias is particularly important when using data acquired outside of a traditional research survey, although research surveys have their built-in biases too. Understanding this is essential for the ethical use of the data, as assumptions might be made without knowing if people or places are missing from the data set simply because of the way it is developed. For example, let's look at the databases the CDDL team scraped for the Ghost Cities project, which Baidu, Amap, and Dianping have curated for the purpose of providing information about points of interest. I like to use the word *curated* because its definition is to "select, organize, and look after items." We often forget that a person or an entity has made selection decisions to create and populate these databases— and decisions arguably *always* create prejudice or bias in the data. Dianping's bias comes from its practice of recording only the businesses it thinks people would specifically come to visit. It may not include, for instance, many smaller businesses that people run out of their apartment to patronize, such as street-side vegetable vendors prevalent in older neighborhoods of Chinese cities. In fact, that bias caused us to originally miscategorize some older residential neighborhoods in the Ghost Cities project, because our model used only Dianping-chosen amenities.

To be sure, data scientists are trained to be aware of biases in data, not only because they can affect results, but also because they can help tell a story from the perspective of the persons or entities who curated the data. Understanding biases that are associated with data is essential for ethical data use. There has been much criticism regarding using data scraped from the web or social media sites because the data is skewed to the people who use these platforms. For example, social media sites such as Foursquare, Twitter, and Flickr are more heavily used in urban environments,[31] and within these areas the majority of data is recorded in commercial areas,[32] which means the data can only be accurately used to describe these places. Self-selection bias can also be manifest in social media because users tend to be young and upwardly mobile; therefore, analysis of this data tells us more about them than it does about a population of retirees on a limited budget.[33] Twitter has a more diverse user population, with more African American and Hispanic users compared to Foursquare, whose users are largely young white males.[34] A user's impulse to contribute data also can introduce a certain level of bias. For example, sending a Tweet might be motivated by wanting to complain about a situation and letting others know. The way data is recorded in the first place can create bias. Corporations often use social media platforms to ask questions about products they want to advertise: they don't care about the answers but simply want to get the word out about something they want to sell—and the results that come from these questions are often hard to interpret.[35]

So while it is clear that data scraped from websites and APIs is biased, this does not mean that its use is off limits for analysis. Traditional research surveys also hold their own bias, as it reflects the interest of the person who asks the questions and the people willing to answer them. This doesn't mean we can't use big data but we must understand these biases to work with data ethically—and to make sure everyone's voice is heard.

Bias in Getty Images Data Explains How Cities Create Buzz

Biases in data can also seep into an analysis because data often reflects how its curators see the world. A telling example from my own work is a project in which I use data gathered by accessing the Getty Images database to understand the arts and entertainment world in Los Angeles and New York City (figure 3.12). My collaborator Elizabeth Currid and I named the project Geography of Buzz because by collecting and analyzing data from the Getty Images database we were able to show how media uses place to create "buzz" to sell images and products.[36] At its core, the Getty Images database holds only images that its curators think will sell. Perhaps more importantly it holds images of events that marketers and public relations firms promoted to Getty—places they wanted Getty to capture. Exploring this bias allowed us to discover the geographic preferences marketers and public relations firms use place to market their products. In a way, we took advantage of the data's bias.

Getty Images is the largest purveyor of photographic images in the world. Images on the database are taken either by staff photographers or purchased by Getty for resale. Everything from pictures of Kate Middleton at Wimbledon and Britney Spears's birthday party to the NATO Summit is deposited into Getty's database and logged for reuse. News outlets all over the world purchase these images to include in media coverage of an event. Getty's online portal allows anyone to see their data as thumbnail images with small amounts of metadata (figure 3.13). This metadata lists the topic the event covers, who appears in the image, and the location where the photograph was taken. Our research team developed a "robot" code to scan the site and retrieve metadata for all of its images related to arts and entertainment. We created our own database, consisting of 6,004 events and totaling 309,414 images, from data easily accessible online, by systematically going through each image and programmatically copying its metadata.

Our researcher then tagged the collected data and categorized it according to different parts of the arts and culture industry—museum or gallery exhibitions, movies, music, theater, TV, fashion, and hype (an event for event's sake; not related to one industry) and once categorized it was geocoded. Maps were then made of each category in Los Angeles and New York. Our research team used spatial clustering algorithms to investigate to what extent events clustered in each city (figure 3.14 shows a fashion cluster in New York City). We reasoned these clusters were areas where the sheer density of images could not have happened by chance.

Analysis of the Getty Images metadata shows that the emphasis in the majority of photographs is solidly on big-ticket, high-profile events that attract the attention of the greatest number of consumers, thereby excluding

3.12 Thumbnails and metadata from the Getty Images database. Image provided by Sarah Williams.

3.13 This image shows all the arts and culture images scraped from the Getty Images database. The circles represent the images taken at the same location. Screenshot provided by Sarah Williams, Spatial Information Design Lab.

NEW YORK

VMA Style Villas At Bryant Park Hotel

VMA Style Villas At Bryant Park Hotel
The Fresh Air Fund Celebrate New Payless Fashion
Caravan Hosts StyleLounge Opening Launch Party
Sarah Jessica Parker And Lord & Taylor Celebrates
Grey Goose Entertainment & Sundance Channel Honor; Baryshnikov Arts Center, 37 ARTS building
Isaac Mizrahi Fall 2007 - Front Row
2006 Art Winners At B
Renee Larc For Milan Spring 7 - Back Stage
Tracee Lacc For Milan Spring 2007
Runway Westside Loft
37 ARTS building
Usher Rehearses For Broadway's "Chicago"
Nike Hosts All-Star 25th Anniversary
Oprah Winfrey Opening Of ABC TV Special "L
The 5th Annual RxArt Ball; Splashlight Studios
Victoria's Secret Angels Celebrate America's Numb
Heidi Klum Launches Body By Victoria The Body Bra
"Tweety Designed By Nicky Hilton" To Debut In Ded
Isaac Mizrahi Hosts "Mizrahi 2
Tommy Hilfiger Fall 2
Bell's Delirium At Dry
BET Presents Rip The Runway
The New Stars Of GREASE At Trick Tock
BET Presents Rip The Runway
Akiko Opening 2007
Miley Cyrus Fall 2007 - Runway
BET Presents Rip The Runway
VH1 Hip Hop Honors - Arrivals; Hammerstein Ballroom
Victoria's Secret Launches Sexy Sony With Santos
Stephen Marbury Launches The Starbury Collection
New York Holds "Construction" Exhibition; New York Design Center
DKNY Jeans Presents Blender Magazine's 5th Anniversary; Studio 45
DKNY Jeans Presents Blender Magazine's 5th Anniversary; Studio 450
Heatherette Spring 2007 After Party
Nick Danger Presents "Flirting With Danger"
Ciara & Missy Elliot "Dripping
Nick Cannon Hosts Boost Mobile RockCorps; Midtown Loft
Animal Fair Magazine's 7th Annual "Paws For Style"
Mos In Manhattan
DIFFA's Dinner By Design Gala
The Museum At FIT Presents "Ralph Rucci: The Art Reem Acra Bridal Collection; Prince George Ballroom
Troika Dialog Debut Of Igor Chapurin's Anastasia; Peter White Studio
Matthew Williamson Fall 2007 - Runway
The Museum At FIT Presents "Ralph Rucci: The Art
A Sneak Preview Of "HGTV Design Star"
"The Other Side" Opening At The Tony Shafrazi Gal
M.A.C. Aids Fund Announces New Viva Glam VI Campaign; Cedar Lake
Beyonce Knowles' Birthday
Marc Jacobs Fall 2007 - Runway
Plum Celebrates Albers & Judd Opening At PaceWild
Hip-Hop Summit Action Networks Presents The 4th Annual Action Awards; The Lighthouse
Trans Continental Records Presents Jordan Knig
Irv "Gotti" Lorenzo Celebrates With Universal Motown; Utopia III Yacht; Pier 60
Dave Chappelle Performs
Chris Mazzilli Invites You To Gottin
LawAndre Shapiro Hosts Behind The Business With Mar
The 2006 Snow Ball Chi
Ambrielle Lingerie Launch Party; Chelsea Art Museum
Timberland Holiday Fashion Preview
Gap & Vanity Fair Celebrate The Launch Of "Indivi; Eyebeam
Samsonite Launches 2007 Black Label Collection
Narciso Rodriguez Fall / Winter Collection After
Gap & Vanity Fair Celebrate The Launch Of "Indivi; Eyebeam
Launch Of Iconic Samsonite Black Label Fashion
Kate Nahon & The National Arts Club Unveil The Be
Opening Night Of Atlantic Theater Company's "Spring Awakening"
Victoria's Secret Pink Hosts The Phi Beta Pink Sorority House
Billion Dollar Babes And ELLE Magazine Host VIP
Billion Dollar Babes Shopping Event
Sophia Kokosalaki for Nine West Fitting
Charlotte Ronson & Ann Dexter-Jones Views Fall 2007;
Sandra Bernhard's "Everything Bad & Beautiful" Of
Uncut- The Ray-Ban Wayfarer Sessions; Irving Plaza
Vogue Presents First Look Show And Performance
VOOM HD Networks Presents "VOOM Portraits Robe
Rush Philanthropic Arts Foundations Youth Holiday Party; Irving Plaza
DDCLAB Flagship Boutique Re-Launch Party
Surface Magazine Party
"The Daily Show With Jon Stewart" Celebrates 10 Years; Irving Plaza
The Weinstein Company Presents The Premiere Screening; Regal Cinemas
Sir Richard Branson Announces Headliners And Co
Tim Gunn Hosts VMA Fashion Scholarship Fund Round; Parsons The New School of Design
Max Azria Spring 2007 Dinner And Cocktails
Final Casting Call For "Live Mansion: The Movie Premiere Celebration Of "The Dolby Station
GQ Magazine Celebrates Heineken Premium Light
Purple Fashion Magazine Party At Beatrice Inn
LIFEbeat: The Music Industry Right Aids An
Hot Chip Performs At CMJ Music Fest; Webster Hall
CMJ Music Marathon Presents The Knife; Webster Hall
J/Us Records Showcase
The Domino Green Issue Party
2006 CFDA Awards - Press Room
Brizo & Harpers Bazaar Host A Cocktail Reception
The Hetrick-Martin Institute Hosts A Pre-Screenin
The New York Times Presents The Emerging Artists; Joe's Pub at The Public Theater
Pre-Opening Benefit For The Watermill Center
Donna Karan's Holiday Party; Stephan Weiss Studio
Opening Night Of "Stuff Happens" - Arriv
Opening Night Of "The Treatment" - After Party; Butter
Celerie Kemble & Robert Verdi Host Stephen Dweck"
IFC Films Screens Wordplay - Arrivals
DKNY Jeans & Sony BMG Film Premiers "East Of Hav
JoJo Celebrates Her Seventeen Magazine Prom Cover; Caravan
Charren Calvin Klein Collection Toasts Francisco Costa
DKNY Jeans Presents Bloodline At 401 Projects
HBO Premieres "The Wire" - After Party
Narciso Rodriguez Spring 2007 Footwear Collection & Re
Mikhail Baryshnikov Hosts Inter-Course: Photograp; 401Projects
Mark Seliger's 401 Projects Presents James Nachtw
Opening Night Of My Name Is Rachel Corrie At Minetta
David Bowie & Alan Cumming Invite You To Cel
The Public Theater Presents Satellite
Patrick McDonald & Patricia Field Hosts "The World of Eve Kitten"; Patricia Field Boutique
The Culture Project Presents The Opening Night
Howard TV On Demand Presents "The Stupid Bowl"
Gallery Hanahou Presents The Opening Reception Fo
The Cinema Society & DKNY Jeans Host A Screening
SOB's, Hot 97 & VH1 Soul Present Who's Next Live
Oilily Opens Soho Flagship Store
Reception For "Stan Lee: A Retrospective"
Janet Jackson Hot 97 FM In-Studio Series
BMI's Latin Alternative Showcase
Paul Smith Soho Store Opening
Rhymefests "Plugg City" Mixtape
The Premiere Of "The King"
Napapijri Flagship Store Opening
Avery Storm Birthday Celebration; Kush
Miramax & Elie Tahari Host A Screening Of "The Lo
Elie Tahari Fall 2007 Presentation
Charlotte Ronson Invites You To A Night Of Shoppi
EA & Def Jam Interactive Celebrate The Release Of
Bravo's "Project Runway" Season
Cinema Society & Frederic
Paul Smith Flagship Store Launch
Ashlee Simpson Hosts Victoria Secret
Stella Fall 2007 Collection
John Legend Performs At Bowery Ballroom
Jazz At Lincoln Center's 5th Annual Spring Gala
Ashlee Simpson Hosts Victoria Secret
Benjamin Cho After Party
Sean Lennon In Concert At Bowery Ballroom
Breaking The Band Live Concert At The 5th Annual
UFC Hosts Emerging Designer Party
Spin Magazine & IFC Films Presents The Premiere O
Catherine Malandrino's Newest Concept Of The Soho
3rd Annual New York Comedy Festival
Nine West, VOGUE And Macy's Host Project Front Row; Skylight Studios
Nicole Romano Fall 2007 - Runway
The 25th Anniversary Of Guess
Yellow Fever By Jamison Ernest Model Casting
The Cinema Society Host A Special Screening
Funkast Record Release After Party
Zac Posen Spring 2007 After Party; Soho Grand Dome
Miss Sixty Spring 2007 After Party Celebration At
Alicia Keys & Kerry Brothers Launch "Krucial Keys"
Tribeca/ASCAP Music Lounge At Canal Room - Day 3
Tribeca Cinema Series Hosts A Tea Party For "Miss Potter"; Tribeca Gall
THINKFilm & Diane Von Furstenberg Host A Screenin
Premiere Of Land Of The Bli
The Tribeca Cinema Series Presents "Sound Bites"
Fifth Annual Tribeca Film Festival Awards Night
The Cinema Society & Hugo Boss Present The Premi
The Cinema Society & Calvin Klein Host A Screenin
Exclusive Sneak Peek Of HBO Mobile; Mr. Cho
Chanel Tribeca Film Festival Dinner At The 5th An
The New York Academy Of Art 2006 Tribeca Ball
Naomi Campbell Arrives At The Manhattan Criminal Court -- Has two locations 100 Centre Street and 346 Broadway
Naomi Campbell Appears In Court; Criminal Court Building either 100 Centre Street or 346 Broadway
CMJ Music Marathon Presents The Cardigans; Knitting Factory
The Clipse Performs AT The CMJ Music Fest; Knitting Factory
Vanity Fair Party For The 5th Annual Tribeca Film
Premiere Of The Road To Guantanamo - Dinner
WNYC Presents The Leonard Lopate Show
Mayor Bloomberg & Tribeca Film Festival Co-Founder; New York City Hall
Brian McKnight Celebrates His Latest Album "Ten"; J&R Music and Computer World
Miss Sixty Fall / Winter 2007 - Front Row
Premiere Of "Yo Soy Boricua, Pa' Que Tu Lo Sepas!"; Schimmel Cent
Gladys Knight Appears At J&R Music and Computer World
Manhattan Theatre Club Presents
Best of Broad
Gen Art's Fresh Faces Spring 2007 - Celebrity Gue
Kiss FM 98.7 Presents The 2006 Phenomenal Woman C; Bridgewater on the South Street Seaport
Village Care Of New York Presents 5th Annual Tuli
MTV Presents Making The Band III Final Concert
Exclusive Pepsi Smash Concert; Empire-Fulton Ferry State Par
Fox Searchlight Pictures Premieres "Little Miss S
Nelly Receives Honor At Black Retail Action Group; Cipriani Wall Street
The Cast of Will and Grace Ring The NYSE Closin
Accessories Council Presents The 10th Annual Ace Awards Gala; Cipriani
2006 Australia Day Ball Honors Olivia Newton John
The Leary Firefighter Foundation 6th Annual Benefit: Cipriani restaurant on Wall Street
Fat Joe Speaks At Vocational Foundations Graduat
MAO Magazine Fashion Week Launch Party
LAByrinth Theater Company's 4th Annual Celebrity; Downtown Auditorium
Sound Of Soul Screening & Press Conference AT The
MTV Presents The VMA Kickoff Concert At Battery P

MAJOR EVENTS (600 - 1,000 IMAGES)
SMALL EVENTS (50 - 100 IMAGES)
AVERAGE EVENTS (100 - 200 IMAGES)
MINOR EVENTS (< 50 IMAGES)
BIG EVENTS (300 - 800 IMAGES)

3.14 This map shows the clustering of fashion-related events in New York City, as identified from the Getty Images database. Clustering was strongest on Fifth Avenue, in the West Village, and in Soho, all of which are iconic New York City fashion districts. The Fifth Avenue corridor is particularly associated with fashion— or at least it is portrayed as such in movies and media. Source: Sarah Williams, Spatial Information Design Lab.

FASHION

ART EVENTS

MUSIC EVENTS

THEATER EVENTS

TELEVISION EVENTS

MAPS BY SARAH WILLIAMS AND MINNA NINOVA, SPATIAL INFORMATION DESIGN LAB, COLUMBIA UNIVERSITY

...e maps above show the density of some types of cultural events in Manhattan based on an analysis of photographs taken from Getty Images.

Mapping the Cultural Buzz: How Cool Is That?

By MELENA RYZIK

Apologies to residents of the Lower East Side; Wilmsburg, Brooklyn; and other hipster-centric neighborhoods. You are not as cool as you think, at least according to a new study that seeks to measure what it ...lls "the geography of buzz."

The research, presented in late March at the annual meeting of the Association of American Geographers, ...ates hot spots based on the frequency and draw of ...ltural happenings: film and television screenings, ...ncerts, fashion shows, gallery and theater openings.

The buzziest areas in New York, it finds, are around Lincoln and Rockefeller Centers, and down Broadway from Times Square into SoHo. In Los Angeles the cool stuff happens in Beverly Hills and Hollywood, along the Sunset Strip, not in trendy Silver Lake or Echo Park.

The aim of the study, called "The Geography of Buzz," said Elizabeth Currid, one of its authors, was "to be able to quantify and understand, visually and spatially, how this creative cultural scene really worked."

To find out, Ms. Currid, an assistant professor in the School of Policy, Planning and Development at the

University of Southern California in Los Angeles, and her co-author, Sarah Williams, the director of the Spatial Information Design Lab at Columbia University's Graduate School of Architecture, Planning and Preservation, mined thousands of photographs from Getty Images that chronicled flashy parties and smaller affairs on both coasts for a year, beginning in March 2006. It was not a culturally comprehensive data set, the researchers admit, but a wide-ranging one. And because the photos were for sale, they had to be of events that

Continued on Page 5

3.15 In 2009, we brought the Geography of Buzz project to an academic audience in New York City and then to a broader public via a West Village gallery. A journalist who saw the exhibition there wrote about it for the New York Times. Source: Page proof provided by NYT, April 6, 2009.

small, underground, or up-and-coming artists. As a result, the spatial picture that Getty Images helps to create is focused on artists who are already popular to the mass market. Although the data does not allow us to understand the underground arts scene, it does help us understand how the big players in the art and culture industry use place to brand their products. In New York a large proportion of Getty images were taken at Rockefeller Center, on Madison Avenue, in Times Square, and in the Meatpacking District. In Los Angeles, images taken in Beverly Hills and on Hollywood Boulevard represented an overwhelming portion of the database. These sites are the iconic places we quintessentially associate with their respective cities. This iconic imagery helps sell a New York or LA attitude simply by placing a product in that place. Ultimately the Getty Images database tells us how media uses place to sell product.

After completing our data analytics, we contacted the Getty Images staff to show them the research, get their response, and ensure that the work we did with their data fit within the ethical standards of their company. Our Getty contacts were excited by the research we did, but that was the extent of feedback we received from them. Our work received a lot of press, including a front-page story in the Arts section of the *New York Times*,[37] and, as a result, Getty Images got much positive press from our research. The maps (figure 3.15) reveal that paparazzi, marketers, and media—those not conventionally involved in city development—have unintentionally played a significant role in the establishment of buzz and desirability in city development.

Social Media Data (Biased or Not) Gives the City a Voice

Data captured for one purpose and used for another is messy, as seen with the Getty Images project. It's full of user bias, not to mention the hurdles you must go through to get a complete data set.

If the urban theorist Kevin Lynch (1918–1984) were living today I believe he would have been obsessed with using the thoughts and ideas posted on social media to better understand the city—thereby creating our era's version of *cognitive maps* (also called mental maps). Lynch developed the concept of cognitive mapping as part of his book *The Image of the City*.[38] Central to the practice is the notion that citizens collectively create an image of what is important in the city based on what they remember to draw on a map (figure 3.16): for instance, when people are asked to map out the routes they

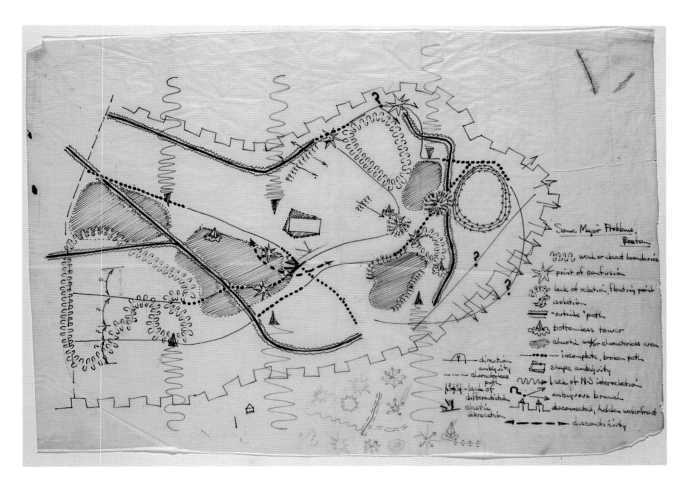

take as they walk to work or to their favorite restaurants, they are instructed to include the landmarks that help guide their way. These might not be landmarks in the traditional sense, but rather more mundane markers, such as the house with the lion statue or a community garden. Creating a map from geo-registered social media data similarly notes people's thoughts in the place where they thought them, which becomes a landscape of feelings, emotions, and personal landmarks.

One of the first projects to download social media and use it this way was a project I worked on in 2011 called "Here Now! Social Media and the Psychological City." In this project I scraped data from Facebook and Foursquare to understand the voice of city inhabitants by reading their

3.16 Key to Kevin Lynch's concept of the cognitive map, popularized in the 1960s and 1970s, is his belief that as we attempt to understand how people perceive the places where they live, memory can tell us a lot about how those places makes them feel and what creates those memories. Source: Kevin Lynch, Some Major Problems, Boston: MIT Libraries, 1960, Perceptual Map, 1960, MC 208, Box 6, MIT Dome, https://dome.mit.edu/handle/1721.3/36515.[39]

geotagged messages on social media. Once downloaded, the data was used to create a map that showed places where users had checked in their on mobile devices to allow friends using the app to locate them (or vice versa) and also check in. On the Foursquare app (figure 3.17), for instance, one user labeled a location "it's creepy you are in my apartment." Other users checked in to places that they labeled with a diverse set of humorous, sad, and lonely descriptions, including "Gay Marriage Equalitocalypse," "Oprah Apocalypse," "Obama's Underpants," "John's Batcave," and "Where Dreams Go to Die." Essentially these maps captured the interests and psychology of city residents (figures 3.18 and 3.19).[40] Since this early project, numerous studies have mined the range of sentiments posted on social media by place and time,[41] whether the users are mapping personal crises to attitudes toward transport and other services.[42] Social media gives us a way to observe the city through the eyes of its inhabitants just as Kevin Lynch sought to do.

Social media data is very good at helping describe qualitative aspects of the city, such as how loud it is or what it smells like, and the results can help

3.17 Foursquare users generate locations where other users can check in. At the time of the Here Now! study, the term apocalypse echoed in many of the names users gave to their locations, including Heatpocalypse, Nicedayocalypse, Overthesnowpocalypse, and TRONpocalypse.
Source: Image created by Sarah Williams, Juan Francisco Saldarriaga, Georgia Bullen, Spatial Information Design Lab

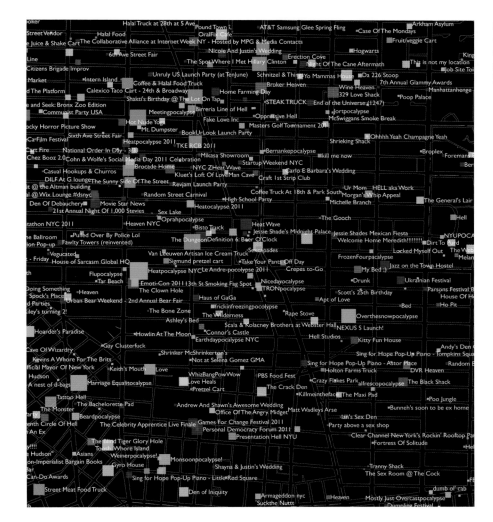

3.18 Map of Foursquare data showing the places people creatively marked on the maps. Source: Image created by Sarah Williams, Spatial Information Desgin Lab

cities make policy. In a recent study, for example, researchers downloaded Flickr data to measure noise levels street by street in London (figure 3.20) and Barcelona; many cities have come to consider noise levels, which have been directly correlated to levels of stress—a public health concern.[43] "Chatty Maps," as the project is called, extracted images from Flickr and uses image processing algorithms to determine the types of noise.[44] Flickr users tag images based on everything from what's in the image ("my best friend") to the event where it was taken ("my best friend's wedding"). To understand noise, Flickr images tagged with noise related terms (e.g., crash, loud, noisy)

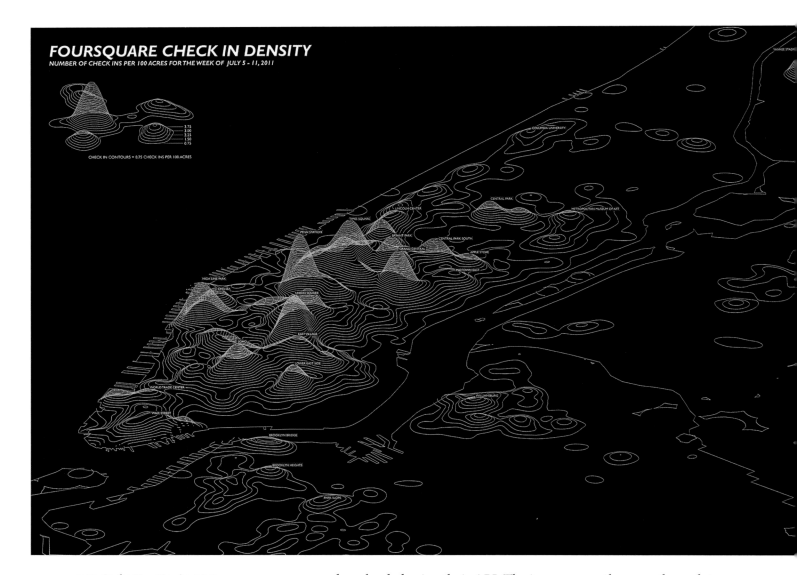

FOURSQUARE CHECK IN DENSITY
NUMBER OF CHECK INS PER 100 ACRES FOR THE WEEK OF JULY 5 - 11, 2011

CHECK IN CONTOURS = 0.75 CHECK INS PER 100 ACRES

3.19 In the Here Now! project we downloaded all the places where people registered while on Foursquare, and then quantified how many times people checked in at those locations. Then we interpolated the values, in much the same way we would have if making a topographic map, but instead creating a topology of social media check-ins. Source: Image created by Sarah Williams and Juan Francisco Saldarriaga, Spatial Information Design Lab

were downloaded using their API. The images were then run through image-processing algorithms, which tag everything in a particular image, to see what might have caused the sound. An image of a building, for example, might return the following tags: building, window, metal, glass, high-rise building, and city reflection. Once the images are tagged this way, they can be used to further identify what the source of the sound might be. The "Chatty Maps" research team gave the following codes: transport, nature, human, mechanical, indoor, and music. These same noise-related Flickr photos extracted from the API were then analyzed for their emotion, using EmoLex (a crowdsourced word-emotion lexicon used for semantic analysis that can estimate the emotion of tags) to analyze user tags, and the resulting analysis

CHAPTER 3

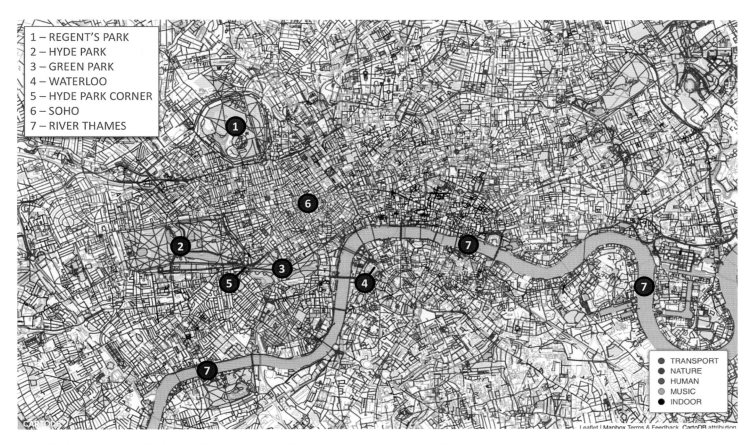

3.20 This image from the "Chatty Maps" study in London illustrates how the authors were able to categorize the types of noises, whether chaotic, calm, monotonous, or vibrant, by running the Flickr images through a visual analytics program.[47]

showed where noise created joy, trust, fear, surprise, sadness, disgust, anger, and anticipation.[45]

The Chatty Maps researchers note that using tags from the Flickr photo does produce some false results: a user might not be uploading an image at the location where it was taken—for example, I might upload pictures of myself in front of the Eiffel Tower after returning to my hotel. These errors, however, did not seem to affect the overall results if large enough data sets were used. The researchers also note that the Chatty Maps project was inspired by a previous study some of them worked on, which developed "smell maps" using a similar method of determining smell words and extracting those words from geotagged social media.[46] The Chatty Maps team publicized its

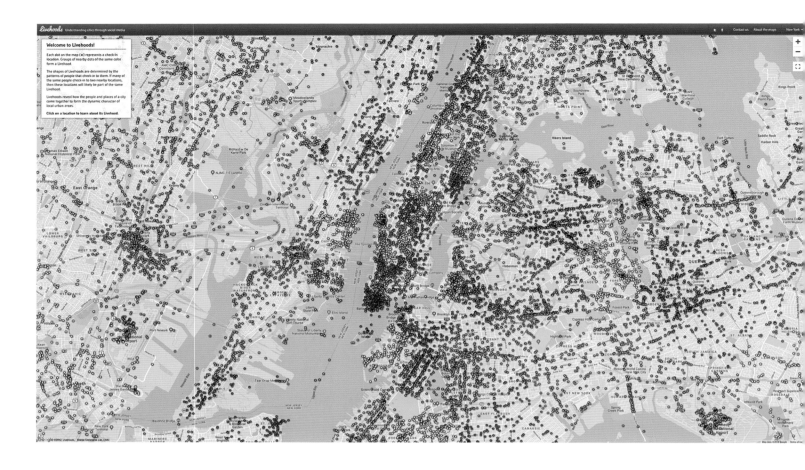

3.21 The map shows neighborhoods as defined by the social and spatial cluster of Twitter users, where their locations are identified by Foursquare's venue data. Source: "Livehoods: New York," accessed August 2, 2019, http://livehoods.org/maps/nyc.

work as a cheap way to get a comprehensive map of sound for the entire city, which ultimately can be used to help urban planners address public health concerns associated with noise levels and mental health.

Understanding emotional responses to the city through social media data can help urban planners learn what people like about cities—but it's essential that urban planners also identify the types of communities that use social media (or do not) in order to know whether whole populations are missing in the analysis. In 2014, Luc Anselin and I set out to describe these in our aptly named Digital Neighborhoods research project. We looked at the spatial clustering of social media data and cross-referenced it with socio-demographic information to describe communities where social media use was prevalent or not.[48] Based on geocoded data we acquired from Twitter and Foursquare firehoses in 2014 (i.e., streams of data those sites sell to third parties), we applied spatial clustering algorithms to the data. We determined that some neighborhoods use digital technologies more than others: social media, for instance, was used more often in commercial areas (all of Manhattan as well as previously or rapidly gentrifying neighborhoods) and less in low-income communities with minorities. The Digital Neighborhoods project actively sought to expose the ethical concerns associated with using data for urban design and planning—by highlighting that not everyone might be "hearing" the messages sent this way. I should acknowledge that had our study included Facebook data our results might have been very different, because Facebook tends to be used by more diverse groups of people across age, race, and class.

A project developed by a team of researchers at Carnegie Mellon set out to test whether they could identify neighborhood boundaries based on data from social media (figure 3.22). They used close to eighteen million records, accessed through Foursquare's API, and linked them to Twitter users' public timeline, which made it possible to determine the location and venues of tweets, and the neighborhoods where users generate them. The researchers then analyzed the data to see not only how the neighborhoods clustered spatially but also to define their respective boundaries; they called the neighborhoods "livehoods" and gave the name Livehoods to the project itself. By interviewing the "livehoods" residents the researchers determined that their results closely matched the residents' perceptions of neighborhood boundaries. For the Livehoods project authors, "the underlying hypothesis of [the] model is that the 'character' of an urban area is defined not just by the types of places found there, but also by the people that choose to make that area part of their daily life."[49] Neighborhoods are notoriously hard to define clearly, depending on and using social media in this way helps to use people's voices to self-identify.

Cities also share crowdsourced data they collect through APIs and open data sites. The call records that come from 311 hotlines, which now exist in most major cities, are a good example. People call these hotlines to report non-emergency events in the city, a large volume of which are public complaints about everything from noise to missed trash pick-up. A spatial analysis (of who calls to complain and who doesn't) can tell a lot about the needs of people who live there. For instance, a study that my research lab performed for the Department of Sanitation in New York City (DSNY) showed that neighborhoods with high numbers of Latin and Asian immigrants didn't appear to call and complain about missed trash pickup (figure 3.22). DSNY knew it was close to impossible that these neighborhoods didn't have the

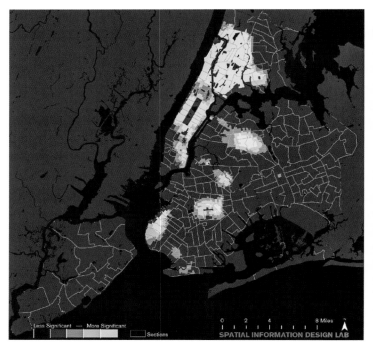

3.22 The map shows hot spots and cold spots of 311 complaints about missed trash pickups. Beyond the wealthy neighborhoods of the Upper East and Upper West Sides, cold spots appeared in neighborhoods with high numbers of Latin American and Asian immigrants. Source: Image created by Sarah Williams, Spatial Information Design Lab

same rate of missed trash pickup than other neighborhoods—so there must have been another reason they were missing in the data. Through interviews conducted by my lab, we found that many residents in these neighborhoods didn't know they could complain about trash service: they were either unaware of the service itself or came from cultures where it's not common to complain about such services. DSNY brought the issue to the person in charge of 311 hotline services, who in turn started an advertising campaign in multiple languages in those communities to help reach populations previously missing from the data, and therefore not receiving the services they needed. Exploring where the data is missing is just as important as exploring where it exists.

Data collected on social media or call hotlines such as 311 help us see the city through the public's eyes, in much the way Kevin Lynch implored. Social media can tell us about the culture of the city from new slang words, such as "Heatpocalypse" did in the Here Now! Social Media and the Psychological City project. At the same time social media data can tell us about the sounds and smells of cities at a scale previously unachievable. These types of qualitative maps could be used to facilitate policy planning. Investigating who is missing from the data sets is just as important as recognizing who is there. In the midst of investigating the 311 data, for instance, New York City officials used the analysis to create action by starting a campaign to provide information about 311 in languages other than English. The Digital Neighborhoods project is an example of how the social–spacial differences that exist among social media users can unintentionally exclude entire populations from analysis when that social media data is analyzed. In the Livehoods project, we found that social media does a pretty good job of helping us self-define our lives—or, in this case, neighborhoods. Social media might provide messy data sets, but it still can offer insights, especially when used by city planners to understand their cities.

The Hubris of Using Data for a Social Good

The availability of new forms of data helped create the field of public health during the First Industrial Revolution, so it's no surprise that public health researchers mined online news media platforms to determine outbreaks of infectious diseases long before social media existed.[50] In fact public health researchers have been leaders in fields that use big data to the benefit of society. Numerous projects have shown the ability of big data to track and predict disease, but this type of analytics, performed without the help of public health experts, often falls short. This is why the Data Action method implores data scientists to include policy experts in any work that is meant to address policy.

One of the early tools used by public health researchers who mined web data was HealthMap.org, launched in 2006 to focus on providing a way to detect an outbreak and to convey the information to the public through a web interface. The interface searches for any news on the web that uses keywords associated with the particular disease in order to create a database of tagged places and topics, which in turn helps to identify locales that might be experiencing an outbreak.[51] In 2010, this technique was applied to Twitter and other social media apps to get a perspective on the cholera outbreak after the 2010 earthquake in Haiti.[52]

Taha Kass-Hout, the former US FDA chief health informatics officer, believes data scraped from social media sites can be applied to help with essential public health concerns such as identifying "non-cooperative disease carriers (Typhoid Marys); [addressing] adaptive vaccination policies; augmenting public health surveillance for early disease detection to create [a] disease situation awareness picture; and updating or enhancing our understanding of the emergence of global epidemics from day-to-day interpersonal interactions, while engaging the public and communicating key public health messages."[53] Kass-Hout co-founded Humanitarian Tracker, a nonprofit organization that "connects and empowers citizens using innovation in technology to support humanitarian causes."[54] Just as with Healthmap.org, much of Humanitarian Tracker's work is dedicated to developing tools that hack into the wealth of information available online to help alert us to possible public health problems.

Social media data and tools like Humanitarian Tracker have been used for monitoring everything from smoking cessation patterns to infectious disease. This method is often referred to as "nowcasting," a term used to describe the ability to predict a range of phenomena happening in the present, near future, or recent past such as the spread of diseases, economic analyses or forecasting,[55] and weather.[56] Aside from its typical use of social media data, nowcasting can also draw from the data *exhaust*, such as search history or logs of purchases on Amazon, generated by interacting with digital technologies. The use of data exhaust has become commonplace in public health research because of its potential for predicting seasonal epidemics such as influenza, which could have impacts on the public.[57]

In 2009, for instance, Google announced it used its vast database of search terms to predict with very high accuracy the number of cases of the flu as well as their locations, claiming that these statistics were comparable to CDC reports for the same period.[58] The data analytics model, called Google Flu Trends (GFT), worked by comparing the pattern of the search term with the known CDC reported physician visits for influenza-like illness (ILI) for the five-year period from 2003 to 2008, and Google made sure to remove seasonally related terms such as high-school basketball.[59] The research was heralded as the next revolution in public health research using big data.[60] But even though Google's initial results proved successful, the algorithms began to fail over time because the model relied too heavily on seasonal search terms and search terms associated with illnesses such as bronchitis and pneumonia. The model also failed because Google didn't update it for changes in types of flu or other factors that could influence the use of the search terms it created.

Google's 2011 self-critique of the failed GFT predictive model admitted that because it did not include cases of summer flu as a baseline, seasonal terms became an overpredictor in the model.[61] Also, the Google team did not update the model over time to reflect new social patterns, such as differences in the age groups that were being affected by new strains of the flu.[62] A detailed analysis of the model in 2013 showed that it was overestimating cases, indeed in some instances showing twice as many cases as compared to the CDC report for the same time period.[63] Another study performed in 2014 on an update of the model Google made in 2014, led by group of researchers at Johns Hopkins with the help of the CDC, showed better performance, particularly in lower income areas. GFT had already received a lot of bad press, however, so no one believed the results, and Google suspended the program in 2015.[64] The project became emblematic of the hubris of big data, with specific critique focusing on the fact that Google did not coordinate closely enough with public health researchers to ensure that its predictive integrity could measure up over time.[65]

Google now favors independent researchers doing their own predictive modeling on flu trends by providing them with quantitative data on search terms. Some believe these new studies are more accurate because they include the specialized knowledge of medical researchers.[66] The problems of the Google Flu Trends model provide a good example of (and an important lesson about) why working with experts is part of the Data Action method: analyzing big data must include experts, in this case public health specialists, in order to understand anomalies in the data and to make it more useful to those professionals.

Looping Back to the Ethics: Data and Disaster Response

How data is applied reflects social norms because the questions we seek to answer and the ways we respond to analysis are driven by the culture we live in. While all the projects discussed so far help to illustrate this idea, contemporary examples from the use of data for disaster relief is perhaps the most illustrative. Disaster response, a field closely linked to public health because of the numerous health concerns associated with natural disasters—cholera, typhoid, and post-traumatic stress among others—has also applied to data scraped from social media to address issues important to its work. This emerging field is often framed as *digital humanitarianism*.[67] Applications and analyses have been deployed to understand natural disasters such as wildfires, floods, earthquakes,[68] and even terrorist attacks,[69] since social media can provide real-time information from those on the ground and responding—and allowing people to know where to respond. The work is similar to the "nowcasting" described earlier in the chapter, as it helps identify the problems associated with a disaster. But rather than being predictive, disaster response tells us what is actually happening on the ground based on messages interpreted by those affected. Jakob Rogstadius, who studies and develops disaster systems for IBM, categorizes disaster response systems that use social media in three ways: First, those that address disaster management directly, sending information through social media about rescue or evacuation efforts. Second, crowd-enabled reporting, which refers to projects similar to Ushahidi (mentioned in chapter 2), where an interface is created to show data posted on social media and news outlets about an event while also allowing users to add their own reports; the data is then made available for anyone to use. And third, automated information extraction using computational tools to determine what is going on during an event by collecting and analyzing reports in the news or in tweets.[70]

This type of digital humanitarianism is not without criticism. The strategies employed often replicate traditional media response to events when people's attention is captured during the event itself but wanes with the news cycle, leaving people to feel stranded or abandoned.[71] Kate Crawford, a researcher who asks questions about the ethical uses of big data, raises three salient points about the limits of crisis data. First, the use of Twitter data during disasters risks missing the experience of people who do not have access to the internet during a disaster, either because the internet is not available in the region itself or because they don't have access to the internet owing to their rural location or poverty. Second, the people who *do* tweet from within the disaster have no control over how their thoughts, reports, and concerns are used or who ultimately benefits from them, which creates a concern for the personal privacy of those involved in the disaster. For example, they may find their tweet widely distributed on the web or through news organizations without their explicit permission. And third, the messages most commonly retweeted during disaster events are often the more sensational stories, which offer a skewed, reconstructed "imaginary of disasters." In other words, people often reconstruct them completely outside the context of the disaster, making it hard to know what the real situation on the ground is, and whether the story as reported is accurate.[72]

Numerous articles cite problems in the digital humanitarian response to the earthquake in Haiti, including everything from data bias to privacy concerns about using people's tweets in ways they were not originally intended to be used, which may be a result of the fact that it was an early experiment in using data this way. A great deal of criticism has been leveled at the response to the Haiti earthquake for not having a structural mechanism in place to follow up on reports distributed via social media. In other words, many people waited indefinitely for help that never came.[73]

Nevertheless, the Haiti earthquake provides numerous lessons for using data in disasters—perhaps most importantly that collaboration among responders, data scientists, and people of the community is essential for addressing consequential ethical gaps in the application of the data set. It's essential that we build teams to work with Data for Policy change: tech experts might be able to build a way to communicate about needs during a disaster, but if there is no way to respond to those needs it can be a greater failure to the public who expects to receive it. Disaster relief systems should be built with people who can respond.

How "White Hacks" Use Data Ethically and Responsibly

The projects I describe in this chapter show the variety of ways people creatively access data that is freely available on the web and social media. Perhaps more important than the means of access are the creative analytics researchers use to gain insights from data. A diverse set of actors—public health officials, disaster responders, transportation analysts, academics, media studies researchers, ethnographers, and people in the private sector—have all used this type of data to address issues they believe have a public benefit. Some important themes and perhaps cautionary tales are exposed as we look at this work, especially the fact that all data has some bias. But data scraped from the web and social media sites might have more bias than most—which means it is essential to develop strategies to critique the resulting insights. The Ghost Cities project provides an example of how we can do this. Data analysts must ground-truth results of data analytics—and by that I mean not only check if our model correctly identified the phenomena we wanted to observe, but also ask the people *on the ground* who are affected by the results of the model we use to confirm that what we identified makes sense. This is a Data Action principle. It means engaging the public, policy analysts, academics, and data scientists to review and critique our analyses to help improve our data model, making sure the voice of the public is incorporated in our results.

A key lesson of this chapter is to be aware of who curated the data you use, and for what reason(s). For example, Dianping's data only includes formal businesses, so many small and informal businesses—often run on the sidewalks or out of someone's living room—are missing from its data set. Similarly, the images in the Getty databases are collected because of Getty's interests in media markets. Data on social media sites such Foursquare, Facebook, and Twitter, each of which has its own distinct user base profile, represent the social norms of their users and what they are willing to contribute. Despite these imbalances and biases, we can still use social media data responsibly to expose underlying patterns, but it is important to remember to develop questions that minimize the data's biases.

Another key lesson of this chapter is the Data Action principle of working with policy experts, which is so central to working ethically and responsibly. The Google Flu Trends (GFT) example amply demonstrates this principle. As I noted earlier, Google's attempt to track flu outbreaks and clusters based on users' search patterns failed spectacularly. In many ways GFT represents the hubris and arrogance of data scientists who do not consider it important and informative to involve subject experts and practicing specialists throughout their projects. This often renders data analytics and analyses needlessly naive and uninformed. Early analysis of the GFT algorithms did include some input from policy experts, most of which was about cross-referencing data with official CDC data, but these relationships were not maintained and did not continue. As a result the model fell out of date quickly because new strains of the flu promoted different search terms that had not been included in the original model, and the model was not updated to reflect ongoing changes in public health literature. The input and ground-truthing of stakeholders needs continuous updating.

Researchers should be cautious about using data scraped from Twitter during disaster events without consulting policy experts who can help coordinate on-the-ground assistance. The momentary, ephemeral nature of tweets about disasters are often sensationalized by the media, skewing the propagation of certain types of tweets that might not reflect the reality of victims' needs.

Working with data responsibly also means proactively seeking out data voids—missing data means missing people. These voids reflect not only those who are left out of data sets but may often indicate biases that contribute to relegating some groups to society's margins. Just as data can guide us to housing vacancies in China, a lack of data can tell who is missing from the picture. The investigation of the 311 data showed that many immigrant communities didn't know they could use a hotline to report a lack of sanitation services. Indeed, some of the stakeholders we need to include may be precisely those hiding in a data void—those who, for a whole host of reasons, turn out not to be among the counted.

Finally, too, privacy is an issue. This happens most insidiously when the data offered up for one reason, with one intention, is suddenly integrated into another calculus and used for a wholly other, different reason. The Psychological City project marks the location of users in their homes, workplaces, or a sacred space—and places their thoughts and emotions on a map that exposes them in ways they could not have conceived and likely would not have imagined. Sure, people agreed to give up some information, and they agreed to the terms of service. But is it ethical for data scientists to use people's data in this way? We know private companies are already using our data in the background to help offer us products, but how do we sort out these questions when our purpose is the public good?[74] What are the ethics that must be considered in these situations?

This brings us full circle to the notion of hacking. There are white hats, the hackers who use their skills for the benefit of society. And there are the black hats, those who use their skills toward endeavors with the intention of purposeful malice or harm. The gray hats lie in the murky waters between them, often believing they are breaking with ethics for the greater good of society. We might all strive to be white hats following the principles of the Hacker Code of Ethics: sharing, openness, decentralization, free access to computers, and world improvements. And yet, to be true white hats, we must always evaluate the possible ways we can misuse data even when approaching our work with the best intentions. To do that we must ask all of those who have a stake in the data analytics we perform to be included in the process. This allows them to evaluate our results to improve the quality of the data and make the analysis more responsive to their needs, thereby helping to create more informed policy change. Unfortunately, data analysis often falls in the space between white and black hats, and it's up to us to interrogate those projects.

SHARE IT! COMMUNICATING DATA INSIGHTS

James Madison, the fourth US president and renowned Father of the Bill of Rights, wrote: "A popular government, without popular information, or the means of acquiring it, is but a Prologue to a Farce or a Tragedy; or perhaps both. Knowledge will forever govern ignorance: And a people who mean to be their own Governors, must arm themselves with the power which knowledge gives."[1] Madison knew knowledge must be shared with the public in order for that public to make educated decisions in a democracy.

Sharing data does so much more than provide access to information. It creates trusting relationships, changes power dynamics, teaches us about policies, fosters debate, and helps to generate collaborative knowledge sharing, all of which are essential to building strong, deliberative communities. Yet interpreting data in its "raw" form can be difficult for the average person; most people do not have the skills to parse through data whether big or small. It follows that the insights found in data must be communicated so that anyone can understand them. How data is delivered affects how it performs in society.

Sharing data is part of our everyday life. Whether that sharing is intentional or not, given or received, it transforms how we understand and *see* the world. We can all probably think of ways that data shared with us changed the way we looked at a topic or idea. Advertisers know this well; they often employ data to sell us everything from the latest fad diet to the most effective toothpaste. In fact, data is used as evidence to drive most anything, including policy. What's so powerful about using data to sell ideas is its easy and seamless association with fact, which gives it legitimacy. That is why data's power needs to be wielded carefully: there is a fine line between using data to improve society and doing it harm.

In this chapter we look at the various ways that sharing data can provide insights to create policy action—focusing largely on sharing that happens through design communication such as data visualizations and mapping. The examples presented show that how we share data inherently constructs power relationships between those who collect, analyze, and present that data and those who consume the results. This chapter calls not only for governments to post their data online but also for data specialists everywhere to develop ways to share and communicate their results so that broad audiences can understand their work. Multimedia, visualizations, images, videos, art, performances, or interactive web pages help non-specialists to read the complexity of data without asking everyone to be a data scientist.

It is hard to believe that data visualizations such as pie charts, graphs, and histograms are a relatively new invention, when compared with maps. They appear on the scene in the eighteenth century and instantly become essential to communicating data. It is said that William Playfair's *Commercial and Political Atlas and Statistical Breviary* (1786) (figure 4.1) was the first of its kind, composed of graphs and pie charts, not maps that summarized the trade between England and other countries. Previously this sort of data had been printed as text. In a recent reprint of the Playfair *Atlas*, the editors discuss the brilliance of the development of Playfair's visualizations: "Graphs convey comparative information in ways that no tables of numbers or written accounts ever could. Trends, differences, and associations are seen at the blink of an eye. The eye perceives instantly what the brain would take second or minutes to infer from the table of numbers, and this is what makes graphs so attractive to scientists, business persons, and many others."[2] Playfair knew well that the tables he developed conferred a sense of legitimacy upon the message, and he used them to argue his position.[3]

Our media culture suits the consumption of data that can be seen in the "blink of an eye," as Playfair's early representations allowed. Data visualizations are excellent for telling stories and generating policy debates because they are bite-size, consumable thoughts.[4] The consumption of news and information has shifted toward using small, discrete visual packages posted on social media sites and blogs.[5] Data visualizations fit social media well because they create visual snapshots of complex ideas. Sharing images

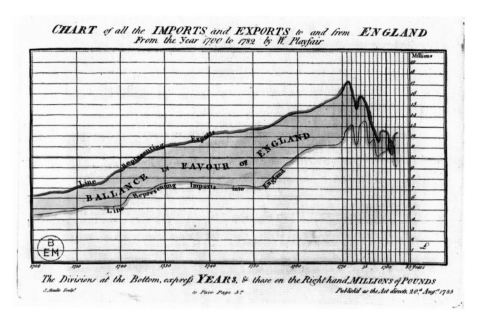

4.1 Chart from William Playfair, *Commercial and Political Atlas and Statistical Breviary* (1786). This visualization is considered the first graph, and Playfair used it to show England's economic strength.

on sites like Twitter, Facebook, and Instagram allows ideas to spread widely while also allowing users of these sites to start to debate the ideas they present.

When data is shared for a specific purpose or intent, it is extremely powerful. That is why sharing data can be so controversial. Opening up raw data for anyone to use is important for transparency among groups and is a symbol of trust. At the same time it holds numerous risks such as privacy concerns, liable errors, and the exposure of underlying social patterns, which the person sharing the data might not have considered. Data shared in visual forms, such as maps, charts, and graphs, hold similar risks as we saw in the first chapter of this book. Highlighting populations on the margins of society can signal them for oppression, as evident in many of the maps developed by technocratic planners. Yet data visualizations, such as John Snow's now-famous map of cholera, can also expose poor living conditions and help to create policies that can improve the public's quality of life (see figure 1.6).

It is important to remember that data is a medium to construct and convey ideas, just as a collection of words makes a story, or an artist who uses paint provides an image of the world. Like words on a page or brushstrokes on a canvas, the message that is shared through data visualization represents the thoughts and ideas of the person who shares it. Whether it's sharing "raw" data through an open data portal, or synthesized into visuals or an interactive website, data is shared with us through the lens of the person, group, or organization that provides it. Even in 1786, Playfair knew well that data is never unbiased and that the public rarely critiques its imperfections, making it a good tool for persuasion.

Constructing Change with Data

When I was eleven, I walked into my social studies class and on the wall saw the Brookes Slave Ship Map (figure 4.2). The week's lesson was on the abolition of slavery. The map illustrated how many enslaved people could fit on a typical slave ship carrying them from Africa to the West Indies. In an initial second, the graphic depiction conveyed the horrors of the slave trade—and that was the point: Where did they go to the bathroom? Did they see the light of day? Did they lie in their excrement? What if one of them died? Who would the corpse be next to? For how many months? How were they given food, lying so close together? Developed by British abolitionists in 1788 as a tool to argue against the slave trade, the map depicted the new standards for slave passage set in England's Regulated Slave Trade Act of 1788. The new regulations were an attempt to control the numbers of slaves allowed on ships, by specifying a space of $6' \times 1'4''$ for each man; $5'10'' \times 1'4''$ for each woman; and $5' \times 1'2''$ for each child. These maps showed that such an upgrade in space was still quite devastating.[6]

The Brookes Slave Ship Map of 1788, as the image came to be known, was quite effective when used to advocate for the abolition of slavery, and was widely reprinted. There is no denying the atrocities once you've seen this data visualization. It creates an emotional response. Its impact comes from its artifice and the narrative that the British abolitionists created around it. The map was developed by the Plymouth Chapter of the Society for Effecting the Abolition of the Slave Trade, and when their London counterpart saw the map they took it upon themselves to distribute eight thousand copies to hang on the walls of pubs throughout the country. The map went viral and was printed in pamphlets, newspapers, media, books, magazines, and so on.[7] Spreading the message of the maps was the abolitionists' intent. They used the maps as an awareness campaign about the horrors of slavery, and they were successful. The work was also included in Thomas Clarkson's *Abstract of Evidence*, a report delivered to the British House of Commons in 1790 and 1791, which helped to create the case for abolishing slavery.[8]

Although the image is said to be the most frequently used representation of the horrors of the slave trade, some groups find its use as an educational tool offensive. These critics see the map as minimizing the humanity of enslaved human beings, rendering them instead as inanimate objects lying and waiting to be saved by the inherently privileged white viewers, who were often complicit in the slavery system, and whom the maps were meant to persuade.[9] They argue that the design itself creates a kind of neutrality between the person viewing the map and what the map depicts, allowing viewers to remove themselves from the picture while allowing them to perceive themselves as saviors. I would argue, however, that this criticism is part of the map's artifice, and if it had been confrontational against the white majority, it might not have been as effective at communicating its message. Showing the horrors of the ship allowed those who viewed the map to interpret its message and create their own meaning from them.

I bring up the much-discussed Brookes Slave Ship map here because it demonstrates the intentionality of creating such images. It also illustrates how differently the image can be read, which helps explain why we must consider the potential ways that the data we share could be used to do harm (see chapter 3). There is no question the Brookes Slave Ship Map was an effective tool. It was made by relatively powerful people to persuade those in governmental power to abolish slavery, and it worked. The criticism of the map by those who see it as reinforcing dominant power narratives may be warranted. Are its critics arguing that the map should not be used? Absolutely not, but they are reminding us that we are not getting the whole story of how and why these maps were created. Their critiques remind us that even though the maps did do some good in the world, they also did so from a position of power.

How do we juxtapose the bias in maps and data visualizations with the good they were able to do? As the child seeing the map for the first time in my social studies class, it taught me a lot in one instant, but it was the narrative my teacher provided around the abolitionist movement that helped me to understand its power. Data visualizations help create a narrative around an idea, and it's the narrative that ultimately has the ability to change people's hearts and minds. When using data for action, we must focus on the story we

DESCRIPTION OF A SLAVE SHIP.

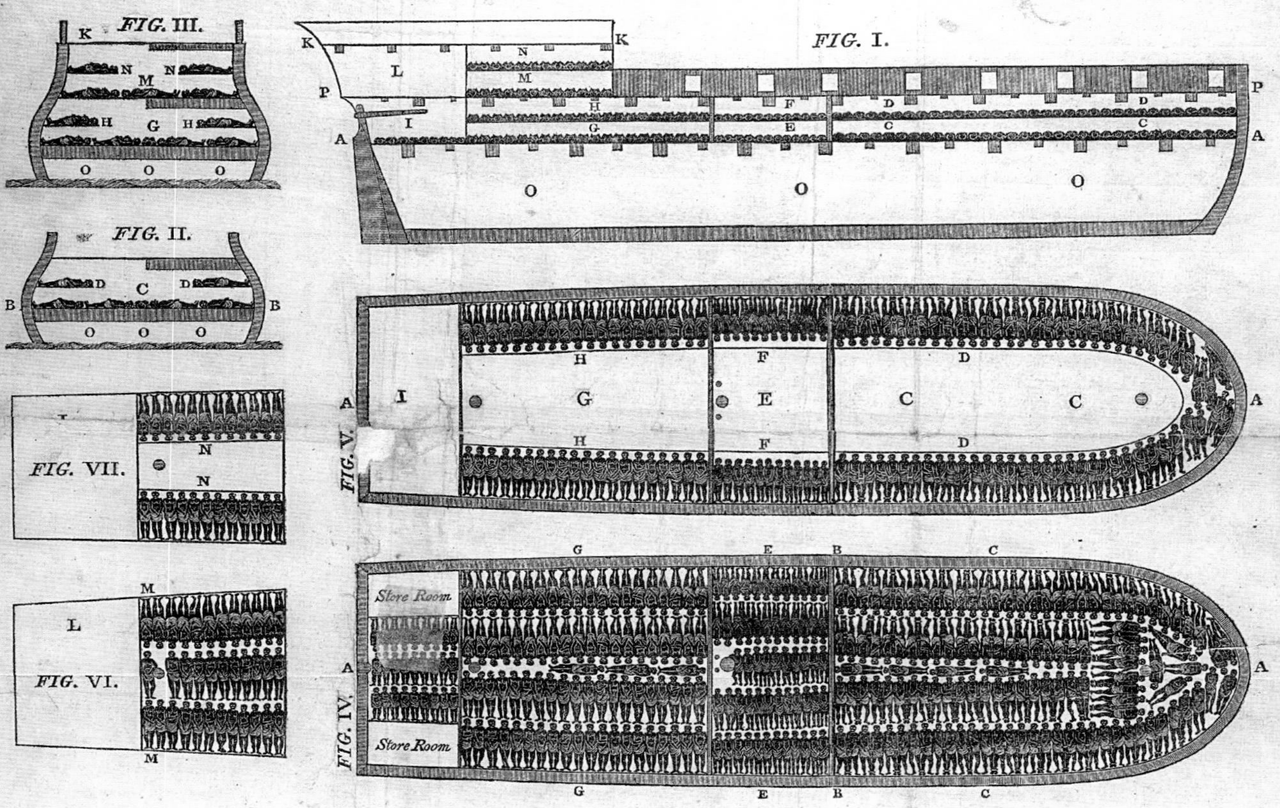
FIG. I.
FIG. II.
FIG. III.
FIG. IV.
FIG. V.
FIG. VI.
FIG. VII.
Store Room
Store Room

The PLAN and SECTIONS annexed exhibit a ſlave ſhip with the ſlaves ſtowed.* In order to give a repreſentation of the trade againſt which no complaint of exaggeration could be brought by thoſe concerned in it, the *Brooks* is here deſcribed, a ſhip well known in the trade, and the firſt mentioned in the report delivered to the Houſe of Commons laſt year by Captain Parrey, who was ſent to Liverpool by Government to take the dimenſions of the ſhips employed in the African ſlave trade from that port. Theſe plans and ſections are on a ſcale of the 8th of an inch to a foot.

DIMENSIONS OF THE SHIP.

	Feet	Inches
Length of the *Lower Deck*, gratings, bulk-heads, included at AA	100	0
Breadth of *Beam* on the *Lower Deck* inſide, BB	25	4
Depth of *Hold*, OOO from cieling to cieling	10	0
Height between decks from deck to deck	5	8
Length of the *Mens Room*, CC on the lower deck	46	0
Breadth of the *Mens Room*, CC on the lower deck	25	4
Length of the *Platforms*, DD in the mens room	46	0
Breadth of the *Platforms* in mens rooms on each ſide	6	0
Length of the *Boys Room*, EE	13	9
Breadth of the *Boys Room*	25	0
Breadth of *Platforms*, FF in boys room	6	0
Length of *Womens Room*, GG	28	6
Breadth of *Womens Room*	23	6
Length of *Platforms*, HH in womens room	28	6
Breadth of *Platforms* in womens room	6	0
Length of the *Gun Room*, II on the lower deck	10	6
Breadth of the *Gun Room* on the lower deck	12	0
Length of the *Quarter Deck*, KK	33	6
Breadth of *Quarter Deck*	19	6
Length of the *Cabin*, LL	14	0
Height of the *Cabin*	6	2
Length of the *Half Deck*, MM	16	6
Height of the *Half Deck*	6	2
Length of the *Platforms*, NN on the half deck	16	6
Breadth of the *Platforms* on the half deck	6	0
Upper deck, PP		

Nominal tonnage — —	297
Suppoſed tonnage by meaſurement	320
Number of ſeamen —	45

The number of ſlaves which this veſſel actually carried appears from the accounts given to Capt. Parrey by the ſlave-merchants themſelves as follows:

Men — —	351	Total 609
Women — —	127	
Boys — —	90	
Girls — —	41	

The room allowed to each deſcription of ſlaves in this plan is:

To the Men 6 feet by 1 foot 4 inches.
Women 5 feet 10 in. by 1 foot 4 in.
Boys 5 feet by 1 foot 2 in.
Girls 4 feet 6 in. by 1 foot.

* This is the uſual manner of placing the ſlaves, but it varies according to the poſition of the ſhip, and the practice of different commanders.

With this allowance of room the utmoſt number that can be ſtowed in a veſſel of the dimenſion of the *Brooks*, is as follows, (being the number exhibited in the plan) and is leſs than 1½ to a ton, viz. †

		On the Plan.	Actually carried.
Men—on the lower deck, at CC	124	190	351
Ditto on the platform of ditto, CC DD	66		
Boys—lower deck EE — —	46	70	90
Ditto—platform FF — —	24		
Women—lower deck, GG —	83	183	127
Ditto—platform, HH — —	40		
Women Half deck, MM —	36		
Ditto Platform ditto, NN —	24		
Girls Gun room, II — —		27	41
General total		470	609

The principal difference is in the *men*. It muſt be obſerved, that the *men*, from whom only inſurrections are to be feared, are kept continually in irons, and muſt be ſtowed in the room allotted for them, which is of a more ſecure conſtruction than the reſt.

In this ſhip the number of men actually carried was ——	351
The number of men ſtated in the plan at 1 foot 4 inches each	190
Difference —	161

As the ſhip on this plan would ſtow 42 women boys and girls in the places here allotted them more than ſhe did carry, ſuppoſing that number of men taken from the mens room, and placed in their ſtead, this will reduce the number of men to 309 in the mens room; of courſe the room allowed them, inſtead of being 16 inches as in the plan, was in reality only 10 inches each; but if the whole number 351 were ſtowed in the mens room, they had only 9 inches each to lay in.

The men therefore, inſtead of lying on their backs, were placed, as is uſual, in full ſhips, on their ſides, or on each other. In which laſt ſituation they are not unfrequently found dead in the morning.

The longitudinal ſection, fig. I. ſhews the manner in which the ſlaves were placed on all the decks and platforms, which is alſo further illuſtrated by the tranſverſe ſections, fig. II. & III. By which it appears, that the height between the decks is 5 feet 8 inches, which, allowing 2 inches for the platform and its bearers, makes the height between the decks and the platform 2 feet 9 inches; but the beams and their knees, with the carlings, taking 4 inches on an average, this ſpace is unequally divided, and above or under the platforms cannot be eſtimated at more than 2 feet 7 inches; ſo that the ſlaves cannot, when placed either on or under the platform, relieve themſelves by ſitting up; the very ſhort ones excepted, nor can *they*, except on board the larger veſſels. The average of 9 veſſels meaſured by Captain Parrey, being moſtly large ſhips, was only 5 feet 2 inches. The height of the Venus between decks was 4 feet 2 inches; of the Kitty, 4 feet 4 inches, both of which had platforms. In theſe ſmaller veſſels therefore they they have not 2 feet under or upon the platforms.

In fig. 1. under the upper deck PP, and the lower deck AA, the beams and the intervening carlings are repreſented by ſhaded ſquares. The beams are alſo introduced on one ſide of the tranſverſe ſections II and III, in order to ſhew the ſpace which a ſlave placed under a beam has to lie and breathe in.

† It muſt be noted, that every poſſible advantage of ſtowing is allowed in the plan. There are or ought to be in each apartment one or more poopoo tubs; there are alſo ſtanchions to ſupport the platforms and decks; for which no deduction is made; but the deck is ſuppoſed clear of every incumbrance whatever.

It may be expected, from this mode of packing a number of our fellow-creatures, uſed in their own country to a life of eaſe, and from the anguiſh of mind their ſituation muſt neceſſarily create, that many of them fall ſick and die. Inſtances ſometimes occur of horrible mortality. The average is not leſs than 1-5th, or 20 per cent. The half deck is ſometimes appropriated for a ſick birth; but the *men ſlaves* are ſeldom indulged the privilege of being placed there, till there is little hope of recovery. The ſlaves are never allowed the leaſt bedding, either ſick or well; but are ſtowed on the bare boards, from the friction of which, occaſioned by the motion of the ſhip, and their chains, they are frequently much bruiſed; and in ſome caſes the fleſh is rubbed off their ſhoulders, elbows, and hips.

It may not be improper to add a ſhort account of the mode of ſecuring, airing, and exerciſing the ſlaves.

The women and children are not chained, but the men are conſtantly chained two and two; the right leg of one to the left leg of the other, and their hands are ſecured in the ſame manner.

They are brought up on the main deck every day, about eight o'clock, and as each pair aſcend, a ſtrong chain, faſtened by ring-bolts to the deck, is paſſed through their ſhackles; a precaution abſolutely neceſſary to prevent inſurrections.—In this ſtate, if the weather is favourable, they are permitted to remain about one-third part of the twenty-four hours, and during this interval they are fed, and their apartment below is cleaned; but when the weather is bad, even thoſe indulgencies cannot be granted them, and they are only permitted to come up in ſmall companies, of about ten at a time, to be fed, where after remaining a quarter of an hour, each meſs is obliged to give place to the next in rotation.

In *very* bad weather, ſome are unavoidably brought on deck: there being no other method of getting water, proviſions, &c. out of the hold, but by removing thoſe ſlaves who lie on the hatch-ways. The conſequence of this violent change from their rooms, which are inconceivably hot, to the wind and rain, is their being attacked with coughs, ſwellings of the glands of the neck, fevers, and dyſenteries; which are communicated by infection to the other ſlaves, and alſo to the ſailors.

The only exerciſe of the men-ſlaves is their being made to jump in their chains; and this, by the friends of the trade, is called *dancing*.

To perſons unacquainted with the mode of carrying on this ſyſtem of trading in human fleſh, theſe Plans and Sections will appear rather a fiction, than a real repreſentation of a ſlave-ſhip. They will probably object, that there is no room for ſtowing cables, and ſuch other utenſils and ſtores as are uſually placed between decks. In a ſlave ſhip (i. e. a full one) theſe articles are either depoſited in the hold, or piled upon the upper deck; and from thence, in caſe of bad weather, or accidents, no ſmall confuſion is occaſioned.—It may be alſo ſaid, the ſlaves are placed ſo very cloſe, that there is not room for the ſurgeon to viſit and aſſiſt them: The fact is, that when the ſurgeon goes amongſt them, he picks out his way as well as he can, by ſtepping between their legs. He frequently finds it to be impoſſible to afford them that relief which an humane man (and ſuch there are even in this trade) would willingly give them. When attacked with fluxes, their ſituation is ſcarcely to be deſcribed. To give an inſtance, (as related by an eye-witneſs) as it ſerves to convey ſome idea, though a very faint one, of the ſufferings of thoſe unhappy beings whom we wantonly drag from their native country, and doom to perpetual labour and captivity: "Some wet and blowing weather having occaſioned the port-holes to "be ſhut, and the grating to be covered, fluxes and fevers among the "negroes enſued. While they were in this ſituation, my profeſſion "requiring it, I frequently went down among them, till at length their "apartments became ſo extremely hot, as to be only ſufferable for a very "ſhort time. But the exceſſive heat was not the only thing that rendered "their ſituation intolerable. The deck, that is, the floor of their rooms, "was ſo covered with the blood and mucus which had proceeded from "them in conſequence of the flux, that it reſembled a ſlaughter-houſe. "It is not in the power of the human imagination to picture to itſelf "a ſituation more dreadful or diſguſting. Numbers of the ſlaves had "fainted, they were carried upon deck, where ſeveral of them died, and "the reſt were, with difficulty, reſtored. It had nearly proved fatal to "me alſo."*

Another objection which may be ſtated, is, that here no room is allowed for the ſailors hammocks. In ſlave ſhips, while the ſlaves are on board, the ſailors have no other lodging than the bare decks, or (in large ſhips) the tops. From this expoſure, they often are wet for a long time together, the rains in thoſe climates being frequent and extremely heavy. There is in wet weather a tarpawling placed over the gratings: if the ſailors to ſhelter themſelves creep under this, they are expoſed to the noiſome and infectious effluvia which continually exhale from the ſlaves below.

It appeared from the evidence given by the ſlave merchants laſt year before the Houſe of Commons, that the employment of the ſeamen, viz. boating up the rivers after the negroes, guarding them on board, cleanſing the veſſel, &c. is of a nature offenſive and dangerous beyond that of ſeamen in other ſervices, and that the ſmall-pox, meaſles, flux, and other contagious diſorders, are frequent on board theſe ſhips.

It is therefore falſely ſaid by the well-wiſhers to this trade, that the ſuppreſſion of it will deſtroy a great nurſery for ſeamen, and annihilate a very conſiderable ſource of commercial profit.—The Rev. Mr. Clarkſon, in his admirable treatiſe on the Impolicy of the Trade, has proved from the moſt inconteſtable authority, that ſo far from being a nurſery, it has been conſtantly and regularly a grave for our ſeamen; *for that in this traffick only, a greater proportion of men periſh in* ONE *year, than in all the other trades of Great Britain in* TWO *years.*

Beſides the time ſpent on the coaſt to complete their cargoes, which ſometimes laſts ſeveral months, the ſlaves are from ſix to eight weeks on their paſſage from thence to the Weſt-Indies.

Now let any perſon reflect on the ſituation of a number of theſe devoted people, thus managed and thus crammed together, and he muſt think it dreadful, even under every favourable circumſtance of an humane captain, an able ſurgeon, fine weather, and a ſhort paſſage. But when to a long paſſage are added, inhuman treatment, ſcanty and bad proviſions, and rough weather, their condition is miſerable beyond deſcription. So deſtructive is this traffick in ſome circumſtances, particularly in bad weather, when the ſlaves are kept below, and the gratings covered with tarpawlings, that a ſchooner, which carried only 140 ſlaves, meeting with a gale of wind which laſted eighteen hours, no leſs than 50 ſlaves periſhed in that ſmall ſpace of time.

As then the inhumanity of this trade muſt be univerſally admitted and lamented, people would do well to conſider that it does not often fall to the lot of individuals, to have an opportunity of performing ſo important a moral and religious duty, as that of endeavouring to put an end to a practice, which may, without exaggeration, be ſtiled one of the *greateſt evils at this day exiſting upon the earth.*

* Falconbridge's Account of the Slave Trade, page 31.

LONDON: PRINTED BY JAMES PHILLIPS, GEORGE-YARD, LOMBARD-STREET. M,DCC,LXXXIX.

want to tell with the data. This is important to remember when we use data visualizations for policy change—and this point is central to Data Action.

4.2 Brookes Slave Ship Map, 1788. Source: "Diagram of a Slave Ship," 1801, https://www.bl.uk/learning/timeline/item106661.html.

Sharing Data Creates Trust and a Commitment to Place

Sharing data can create civic action, and this chapter's case study shows how sharing data openly, through both "raw" data and visualizations, can change power dynamics and provide an essential resource for communities. Our case study takes us to Nairobi, Kenya, where I have worked since 2006 to improve the conditions for mobility in the city and have successfully used data as a tool in that work. Probably the most important benefit of sharing data with the people of Nairobi is that it created a sense of trust and commitment to *place*. Data standards are very important when sharing data, and by sharing our transport data in a standardized transport format we helped extend its use beyond the project we were working on; so that others could easily pick it up and use it. Visualizing their city's data transformed the way Nairobians saw their transportation system, which comprises an informal network of buses called matatus, and it created a platform for discussion. The visuals became an instant planning tool, helping matatu owners, the government, foreign aid agencies, and the public as well to make more informed decisions about the future of Nairobi. Perhaps most importantly our team recognized that the data we collected in Nairobi represented the people who live there, and by including input from matatu drivers and owners as we developed our visualizations, we hoped the map would accurately represent them. How the map was presented to the public was also important; we did not intend the visualizations to be used to marginalize or criminalize certain forms of mobility but rather to allow them to be more integrated into planning processes.

So, you might be asking, what is Nairobi's narrative around mobility? For starters, the city's biggest planning problem is congestion, which often brings the city to a standstill—it can take hours to travel a few miles. Understanding the root causes of congestion, as well as coming up with strategies to alleviate it, have been part of my work for some time. As early as 2006, I was involved in a project with Columbia University to develop a transportation model for Nairobi, the goal of which was to offer recommendations for ways to alleviate traffic congestion.

This is where our story begins. Obtaining data for our work in Nairobi was challenging, as is true for many rapidly developing cities. My team spent months trying to acquire even the most basic data sets about roads, land use, and population. Even more disheartening was that this data about Nairobi existed thanks to earlier infrastructure projects, but my project team was unable to acquire it. This data, created by the Japan International Cooperation Agency (JICA) for Nairobi's National Spatial Data Infrastructure (NSDI) project, was meant to be openly available so it could be used to improve decision making in the city.[10] Our team struggled to obtain this open data, even though the funds that paid for its development specified that it should be made available via download on Kenya's Spatial Data Portal. A visit to the portal where the street files should have existed yielded a broken link. Eventually we determined the agency responsible for keeping the data; we were told that it would cost $10,000 for one license and that our multiple collaborators, including the Kenya Institute for Public Policy Research and Analysis (KIPPRA), as well as experts from our team at the University of California at Berkeley, would also have to pay. What we discovered was that data developed to be open to everyone was in fact a closed resource. Even though our project was an adequately funded study, we did not have the budget to acquire the information by paying for it.

Navigating the process to acquire data in Nairobi was hard. We seemed to encounter barriers at every turn. Even our local partner KIPPRA hesitated to share the basic survey data they obtained about transportation behaviors in Nairobi. There are many possible reasons for this: perhaps KIPPRA wanted to control the information so they could control the message; or maybe KIPPRA had become weary of the numerous outside actors who participated in the city's development, only to leave a few years later. They may have had concerns about our commitment to Nairobi. So many NGOs and multilaterals invest in projects in the city only to leave without a solid implementation plan. We had yet to create a trusting relationship to show our commitment to Nairobi.

While KIPPRA may not have trusted us, Nairobi's municipal government saw our interest in the data as an opportunity to extract payment. We had to come up with another way to acquire the data, and the best solution was to create it ourselves. Then we would share it for anyone to reuse, as we knew a lot of our colleagues in Kenya and beyond were experiencing the same problem. We enlisted the assistance of students from Columbia University and the University of Nairobi, and we created the first openly available digital data set of roads and land use by digitizing paper maps and satellite images (figure 4.3). This data set was essential for our transportation model; it helped

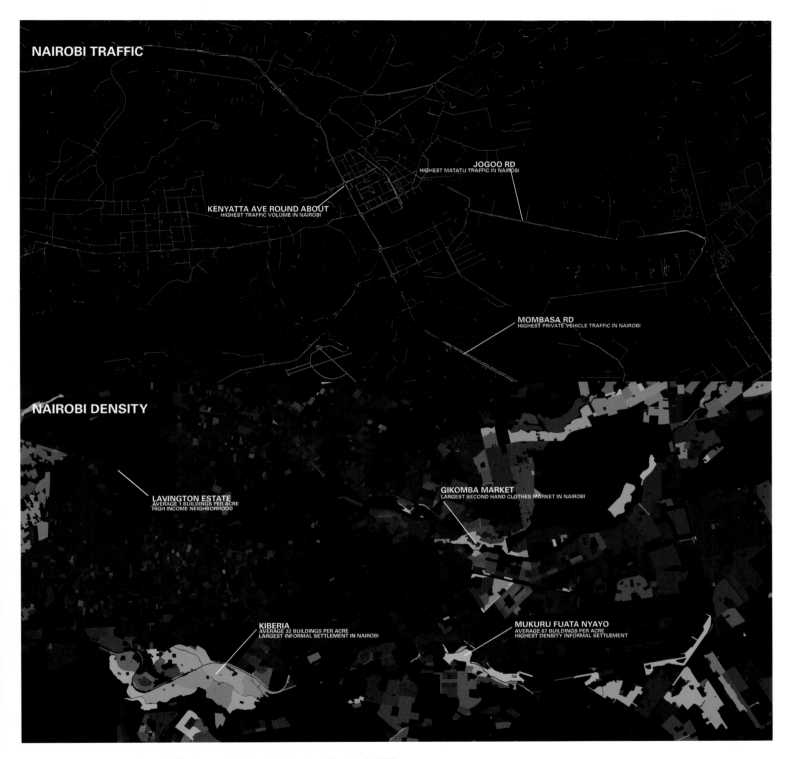

NAIROBI TRAFFIC

JOGOO RD
HIGHEST MATATU TRAFFIC IN NAIROBI

KENYATTA AVE ROUND ABOUT
HIGHEST TRAFFIC VOLUME IN NAIROBI

MOMBASA RD
HIGHEST PRIVATE VEHICLE TRAFFIC IN NAIROBI

NAIROBI DENSITY

LAVINGTON ESTATE
AVERAGE 1 BUILDINGS PER ACRE
HIGH INCOME NEIGHBORHOOD

GIKOMBA MARKET
LARGEST SECOND HAND CLOTHES MARKET IN NAIROBI

KIBERIA
AVERAGE 32 BUILDINGS PER ACRE
LARGEST INFORMAL SETTLEMENT IN NAIROBI

MUKURU FUATA NYAYO
AVERAGE 67 BUILDINGS PER ACRE
HIGHEST DENSITY INFORMAL SETTLEMENT

4.3 The first openly available data set for Nairobi, developed by Sarah Williams
while Co-director of the Spatial Information Design Lab at Columbia University.
Image created by Sarah Williams.

us make recommendations for removing roundabouts because the data models showed they were having severe, negative impacts on the organized flow of traffic.

For many years, the data we created for Nairobi was the only openly available road and land-use file. It was downloaded from our website hundreds of times and even integrated into Google Maps and Open Street Map—the roads seen in both mapping platforms were first derived from our original files. It's amazing that this data is still used today, even though it is quite dated, because it is the only openly available land-use data set for the city.[11] While the data itself was useful to analyze mobility, an even greater benefit was the trusting relationship it allowed us to develop with KIPPRA. Sharing data in this way manifested our commitment to transportation policy reform in Nairobi, and it was seen as a contribution beyond what was typical for outside actors. Sharing data helped create the trust needed to turn data into action.

Making and Sharing an Essential Resource to Create Action

Even though our team created the first publicly available road data set for Nairobi, we were still missing essential data on Nairobi's matatu system—the main form of public transit in the city. An estimated 3.5 million Nairobi citizens depend upon this system every day, and yet at the time there was no publicly and widely available guide to navigating it. The matatu system is made up of privately owned, fourteen- to fifty-six-seat buses. These vehicles are owned by hundreds of different operators who self-regulate the system through the Matatu Owners Association, which helps settle disputes among owners and advocates. The system is loosely based on the former public bus system that fell to privatization in the 1980s. The matatu system sees little government involvement beyond providing licenses to drivers to operate— and from time to time the government creates enforceable mandates, such as banning matatus from the center of the city. Ultimately, however, the government does not want to be responsible for this system, nor to even appear to be in charge of it. Given these politics, the government has had little interest in collecting data about where the matatus go, yet having this information would allow for better transportation planning.

Therefore, our team decided to collect data on the matatu system— not only to provide an essential resource for our transport work but also to create a resource that could be used beyond our project. The results had unprecedented impacts. We began in earnest with a visit to Nairobi's City Hall to identify the location of routes. We found a document that described in general terms some of the matatu routes: the #8 matatu, for example, travels from "Town to Kibera and Junction." The Microsoft Word document was often missing key information about the roads the matatus took along the routes, or where the matatus stopped to pick people up. It became clear that even the average Nairobi resident did not have information about where matatus traveled because there was no transport map of the city. As a result, my colleagues at the University of Nairobi's Center of Computing for Development Lab (C4D Lab), Columbia University's Center for Sustainable Development (CSUD), and the Civic Data Design Lab at MIT applied for a grant from the Rockefeller Foundation to collect geo-registered data on the location of matatu routes and stops that could be used for our model and eventually provide an essential map for both residents and visitors to Nairobi.

Armed with funding from the Rockefeller Foundation, this newly formed American–Kenyan team spent roughly the first eight months developing and testing cell phone software that would collect data while riding the matatus. During this development process, the research team in coordination with KIPPRA held a series of workshops for the local transit community. Workshop participants included members of the government, academia, the head of the Matatus Owners Association, matatu operators and drivers, nongovernmental organizations, and the local technology community. This collaboration succeeded in gaining support from these various actors (see figure 4.5).

Creating data in standardized formats is essential for extending its use beyond the research project for which it was intended. When we began collecting the data in early 2013, we used the General Transit Feed Specification (GTFS) open-data format to store the transit route data. GTFS is a data standard used for transit-routing applications, which allows software to find the best route when given a starting point and a destination. Most readers will be familiar with the Google Maps provision of driving, biking, and public transportation directions, all of which use the GTFS data standard to determine routes. There are numerous free applications that use GTFS data as their base, thereby extending the possibilities for what this particular collection of data can be used for upon release. There is, for example, an open-source application called Open Trip Planner, which is similar to Google

4.4a , 4.4b, and 4.4c University of Nairobi students collect data on the matatu system.
Source: Elizabeth Resor and Adam White, courtesy of Digital Matatus

4.5   Workshop held to engage transportation stakeholders in Nairobi. Source: Sarah Williams

transit directions. But Open Trip Planner is a bit smarter because it has an analysis tool that allows users to determine how long it takes in both time and cost to get to different neighborhoods of the city, showing how accessible they are to transit.

Using GTFS, however, is no simple matter. It might be clear to a data scientist, but it is very abstract to the average person, including the government worker and the matatu driver. In order to make the data understandable and to communicate the matatu system to the broadest of publics, we decided to translate the data into a map that anyone could read and use.[12] Borrowing graphic language from London, Paris, and New York, we designed a stylized map of the Nairobi matatu system, grouping the matatus by the nearest major street and giving each corridor a color (figures 4.6 and 4.7). Major stops and points of interest including parks, airports, and landmarks were added to the map and allowed users to position themselves in Nairobi's geography.[13]

4.6 Final map of the Matatu System, 2014. Source: Digital Matatus

4.7 Process map for the development of the Digital Matatus System map. Source: Sarah Williams, Civic Data Design Lab

"For the First Time, We Can See Our Power"

One of the most powerful moments of the Digital Matatus project occurred when we were able to successfully share the data and maps we created with all of our partners (from government officials to matatu drivers) and the public of Nairobi. When the team presented the stylized maps to the matatu drivers and owners they were excited—for the first time they could see the comprehensive system they had created. The matatu owners are the de facto planners of the city's transportation system—and they instinctively began to use the map to plan new routes for Nairobi's future. The Digital Matatus map also opened the eyes of government officials. Until they saw this map of matatu routes they were uninterested or perhaps didn't understand the need for our data collection project. Once visualized, the map became a powerful tool for their politics. The Ministry of Transport held a press conference releasing our visualization as the official matatu map of the city.[14] This happened because the government trusted the data we had created, and did so in large measure because we had kept officials informed about how the maps developed; all along the way, they trusted the data. The press conference helped create a discussion among the government, the transportation community, and the public about the future of the matatus system. Participants questioned the government about how it would respond to necessary service changes. After the press event, the downloadable maps went viral on social media. Large copies were printed in local newspapers allowing anyone to make use of the data we had collected and the map we had designed to visualize that data.

Our team also held a hackathon at the University of Nairobi to teach the local technology community about the GTFS data format and the open-source software that extends its use (figure 4.8). Our hackathon spawned two mobile applications using the collected data before the paper map was officially released. One of these programs, Ma3Route, became one of the most-used transit applications in Nairobi.[15] Users of Ma3Route share real-time data about the matatu system, noting changes to the routes, traffic

4.8 Hackathon at the University of Nairobi with the local technology community.

CHAPTER 4

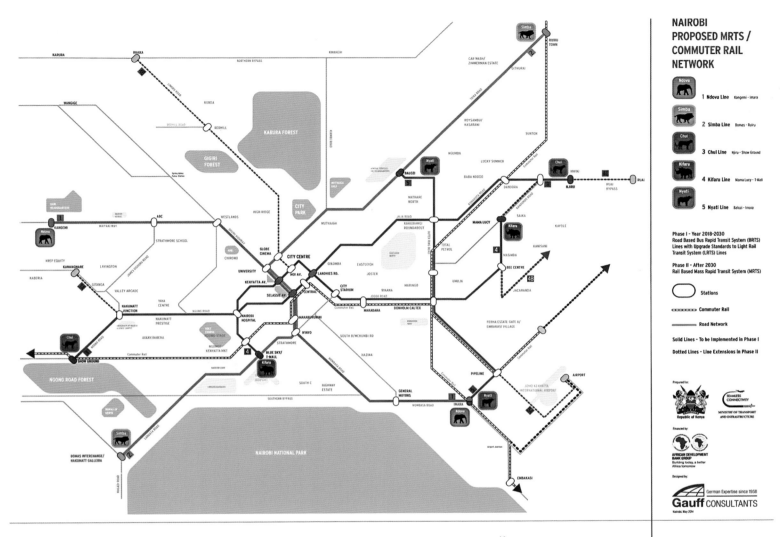

4.9 UN Habitat Map that copied the style of the Digital Matatus Map for a proposed BRT system.[16]

accidents, and traffic congestion. In due course, it won Kenya's Vision 2030 ICT Innovation Award in April 2014.

The Institute for Transportation and Development Policy (ITDP) and consultants at UN Habitat also used Digital Matatus data as the basis of the planning of Bus Rapid Transit (BRT) Service Plans for the city. Nairobi city planners used the data to develop a map of the BRT system that looked like a copy of our map. The city planners were smart to borrow the visual language of our map to earn support for the BRT project because the Digital Matatus map had become an icon in the city (figure 4.9).

4.10 The maps went viral on social media and the internet, and were printed in the local newspaper. This image shows the map centered on a page of Nairobi's newspaper *The Star*. Source: Photo by Sarah Williams.

Digital Matatus: Co-creation Is Powerful

The early work accomplished by KIPPRA paved the way for our now legendary Digital Matatus project because we created a trusting relationship between our research team and Nairobi's transport community. The work of digitizing the matatu routes encompassed many of the methods that I argue are important when employing data for action. Our team built a data set and developed open-source technology to create and edit the data. We shared the data openly by posting it online in a standardized format (to help extend its use beyond our Digital Matatus project) and by creating visualizations. In time, transportation actors in Nairobi used the data to help change some of their city's policies. Sharing data, both as visualizations and in a standardized data format, did indeed extend the life of our transport work in Nairobi and build trust. It allowed others to use the data for their own policy change, create a community around the data, and provide an essential resource to the public. The Digital Matatus project shows that data visualizations are powerful vehicles for generating debate and presenting evidence for planning strategies. The stylized transit maps we developed allowed the government to engage in conversations with the public. Nairobi's matatu operators used the map to identify and develop new routes for the system. Perhaps more importantly, the citizens of Nairobi now have essential information for navigating their city.

Digital Matatus has inspired cities all over the world from Cairo to Bogotá—twenty-six cities in total at the time of this writing in 2019. This network led us to launch a Global Resource Center for the Development of Informal Transit Data with headquarters in Mexico City and Addis Ababa in 2018. The center provides open-source tools, trainings, links to other cities that have done this work, as well as assistance with policy impact and integration. Ultimately the Digital Matatus project has given life to a new form of collecting data on informal systems, one that is collaborative, open, and transparent.

Maps Are a Powerful Medium—They Persuade

Maps are associated with truths and can be powerfully persuasive; however, their presentation of the data can be intentionally or unintentionally misleading, and some people are therefore wary of using them. It can be easily argued that we rarely pay enough attention to the sources on the maps we read, let alone critique their accuracy or the data they are based on. Indeed, as we've seen, the very act of developing data visualizations involves a bias since designers must choose what to simplify and what to abstract to make their maps. Therefore, more critique is warranted.

Critical cartographers including the likes of Brian Harley, Denis Wood, John Pickles, Michael Curry, Jeremy W. Crampton, Sarah Elwood, Annette Kim, and Matthew Edney argue that maps are inherently political; that what is added to and left off a map illustrates or points to social constructs.[17] For example, cartographers often leave out poor areas or slums, alleyways, and much more to display a pristine representation of place that serves their purpose. Studying the symbols on a map can reveal systems of power and control. Critical cartographers ask us to interrogate the political meaning behind maps. Being political doesn't necessarily make the construct of a map harmful—quite the opposite. The politics of maps can be used for good, too.

David Harvey, a Marxist geographer, argues that maps are a tool of power and control because only those in power have the wherewithal to create them, and they often create a one-dimensional perspective on a topic or idea that helps them to retain that control.[18] Brian Harley, a well-known cartographic historian, argues that the practice of developing detailed property records such as cadastral maps is akin to the practices of old-world, agrarian landowners controlling their property. According to Harley, "Accurate, large-scale plans were a means by which land could be more efficiently exploited, by which rent rolls could be increased, and by which legal obligations could be enforced or tenures modified. Supplementing older, written surveys, the map served as a graphic inventory, a codification of information about ownership, tenancy, rentable values, cropping practice, and agricultural potential, enabling capitalist landowners to see their estates as a whole and better to control them."[19] Harley thought this description was fitting for city planners who, he believed, were using their data to divide the modern city into parcels that could be easily bought and controlled.[20] Rather than using the data to improve the quality of life for the citizens, he believed their data-collecting strategies used the cloak of scientific knowledge to advance predetermined objectives. Yet even with his cynicism toward the development of cadastral maps we read about in chapter 1, Harley is not arguing that we should not use maps, but rather that we ought to acknowledge that they are not "value-free."

Making a System Visible to Persuade

Some scholars might claim that by developing a map of Nairobi's matatu system we created a tool of oppression: we made visible an informal system, and the visibility could now be used as a tool to control the matatus with unnecessary regulation or perhaps even cause a complete dismantling of the system. Others might say that in developing a map in a style that mimics European and American representations of transportation, we were framing the data within a colonial construct. While these criticisms might be warranted, matatus have not been made illegal in Nairobi, and the government is still largely uninterested in regulating the vehicles and we have provided an essential resource that the public uses to navigate the city.

Let's take a moment to pick these ideas apart. Why weren't the maps in Nairobi used as a tool for control? I would argue this is because the maps and data were developed collaboratively, and through that effort the maps were co-owned by multiple actors, with none having a singular power over its message. Nairobi's matatu drivers do equate the maps to those of European and American transport systems, and that gives them a sense of pride in their system. Rather than being viewed as part of an informal economy, equating the matatus with formal systems helps show that they contribute an important public infrastructure in the city.

The matatu maps "do no harm" because they tell the untold story of the matatu system. The narrative created by the matatu maps does not paint a picture of an unsafe, corrupt system that needs to be removed but rather one in which transportation actors need to work collaboratively to make the current system more usable. In other words the map using the European representation allowed the matatu driver to be heard. Just as the narrative around the Brookes Slave Ship Map was about abolishing slavery and was meant to speak to those that could change that narrative, the matatu map was to meant to speak to participants and stakeholders both local and foreign, showing that matatu drivers and owners currently run Nairobi's transport system and their voices should be heard. Previously, it was thought that each matatu owner was operating quite independently, and as such any planning decisions that addressed the whole system would be impossible. The Digital Matatus map was shown to those seeking to influence transportation policy in Nairobi, such as the World Bank, to demonstrate that working with the matatu association would be essential for any transportation plan for the city, including the integration of more formalized systems like the Bus Rapid Transit. It is important that we acknowledge maps as a tool for control, and the Digital Matatus project shows how we also used maps to control the narrative around matatus in Nairobi. Our team used the maps as an advocacy tool, representing our ideological positions toward helping those on the margins. There was certainly bias, but does that cause harm? This is the question all data visualizers must ask themselves.

Transforming Map Power for Those on the Margins

The Million Dollar Blocks project transformed the idea of a map from one of power and control into a tool to advocate for those on the margins of society. Our team at the Justice Mapping Center and the Spatial Information Design Lab at Columbia University developed this project to create maps to expose the unconscionable cost of incarcerating people both in terms of actual dollars spent and the cost to the communities the incarcerated come from. Our team transformed a typical crime map, which marks neighborhood hot spots in need of policing, to construct a map identifying neighborhoods in need of investment. Crime maps are often criticized because the way they are used does not address the systemic causes of crime, namely the social problems such as a structural racism, lack of proper access to education, healthcare, and jobs that stem from living in communities of poverty. So while crime might be reduced temporarily in one community because of a police crackdown, it often moves somewhere else. Million Dollar Blocks transforms the typical crime map from a tool of control into a tool to expose the massive amounts of money we spend to incarcerate people rather than address the systemic reasons for their incarceration.

The Million Dollar Blocks maps were developed using prisoner intake data, which includes information about where people lived before they went to prison. These addresses were plotted on a map and cross-referenced with the amount it cost to incarcerate these people, then data was summarized by totaling all the costs for each block. The result created a map showing that in some communities there are concentrations of blocks where more than a million dollars is spent to incarcerate residents (figure 4.11).[21] Often both physically and socially isolated, these neighborhoods typically lack the resources to alleviate the causes of incarceration, including access to education, job training programs, or prison reentry services.[22] The maps conveyed a simple idea: we spend millions of dollars to incarcerate people, so how might we spend that money better in these communities?

The images are at once alarming and captivating because they scale prison policy down to the size of a city block—something that everyone can understand (figure 4.13). Visualizing the data at the scale of the block was an important method for contextualizing the vast sums of money spent on incarceration. The maps show million-dollar blocks as bright red on a black background (a combination rarely seen on maps), which marks the issue as alarming at first glance. Maps showing race and poverty were presented alongside prison spending, allowing those who read the maps to connect incarceration to high levels of poverty and racial segregation. The causes underlying the million-dollar-block designation are multifaceted, of course, and there is no single answer. These maps were developed not to provide a solution, but rather to discuss the issue and present a message: we spend millions of dollars to incarcerate people in the United States, and the cycling of people in and out of prison has become a big business.[23]

CHAPTER 4

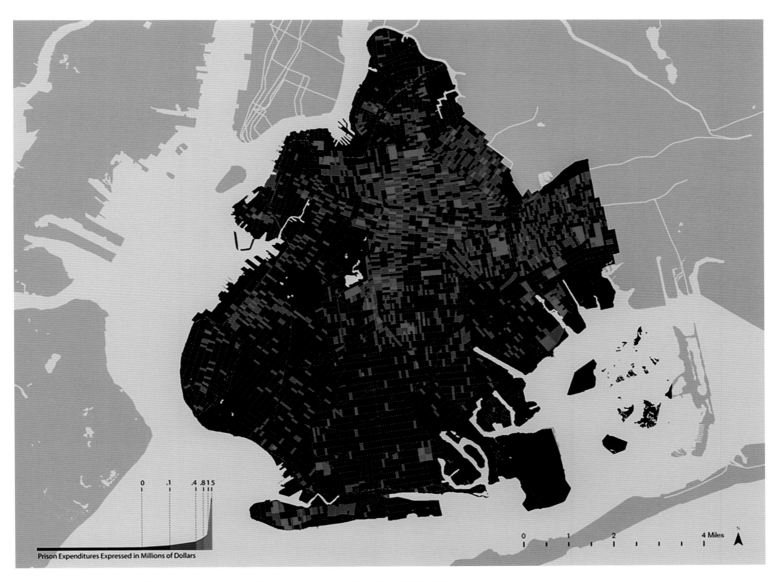

Prison Expenditures Expressed in Millions of Dollars

0 .1 .4 .815

0 1 2 4 Miles

4.11 This Million Dollar Blocks map shows prison expenditures in Brooklyn, NY in 2003. The bright red blocks show where over a million dollars was spent in just one year. *Source*: Maps created by Eric Cadora, Laura Kurgan, David Reinfurt, and Sarah Williams, courtesy of Spatial Information Design Lab.

4.12 This map, seen on the wall of the Design and the Elastic Mind Exhibition at MoMA in 2008, shows where each person incarcerated in 2006 lived and connects their home to the prison they were sent to in upstate New York. The blocks shown in bright red underneath the lines are ones where more than a million dollars was spent to incarcerate residents of that block. The maps on the top compare percentage of people of color, percentage of people in poverty, and prison admissions.
Source: Maps created by Eric Cadora, Laura Kurgan, David Reinfurt, and Sarah Williams, courtesy of Spatial Information Design Lab.

Percent Persons of Color, 2000.

Percent Persons Below Poverty Level, 2000.

Percent Adults Admitted to Prison, 2003.

BROOKLYN COMMUNITY DISTRICTS	% POPULATION	% POVERTY	% ADMISSIONS
BROOKLYN CD 1	6.51 %	9.08 %	5.37 %
BROOKLYN CD 2	4.03 %	3.58 %	4.64 %
BROOKLYN CD 3	5.83 %	8.10 %	16.51 %
BROOKLYN CD 4	4.24 %	6.34 %	9.34 %
BROOKLYN CD 5	7.04 %	9.30 %	14.45 %
BROOKLYN CD 6	4.23 %	2.60 %	3.08 %
BROOKLYN CD 7	5.02 %	5.03 %	3.82 %
BROOKLYN CD 8	3.78 %	4.15 %	9.46 %
BROOKLYN CD 9	4.26 %	4.14 %	4.43 %
BROOKLYN CD 10	4.96 %	2.79 %	0.91 %
BROOKLYN CD 11	7.05 %	5.54 %	1.35 %
BROOKLYN CD 12	7.39 %	8.54 %	1.32 %
BROOKLYN CD 13	4.23 %	4.94 %	3.41 %
BROOKLYN CD 14	6.76 %	6.22 %	3.79 %
BROOKLYN CD 15	6.48 %	4.42 %	1.20 %
BROOKLYN CD 16	3.48 %	5.92 %	8.43 %
BROOKLYN CD 17	6.75 %	5.40 %	5.29 %
BROOKLYN CD 18	7.96 %	3.91 %	3.20 %
BOROUGH TOTAL	100.00 %	100.00 %	100.00 %

Comparisons Expressed as Percent of Borough Total.

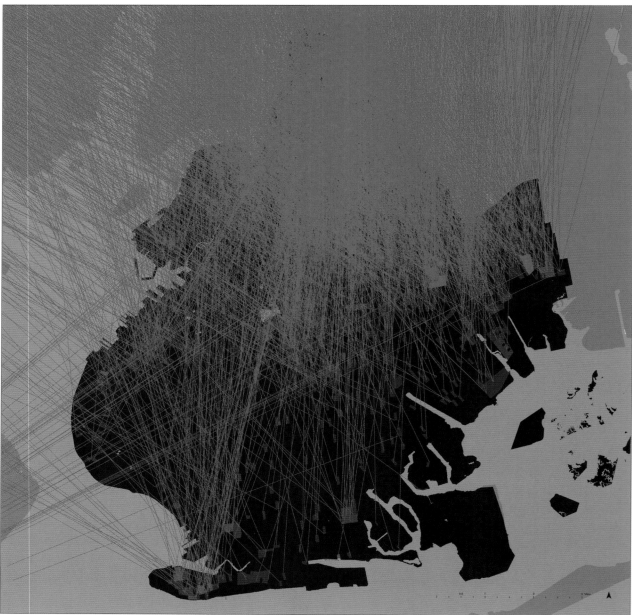

Prisoner Migration with Expenditures by Block, 2003.

BROOKLYN, NEW YORK CITY

ADDED UP BLOCK BY BLOCK, IT COST $359 MILLION DOLLARS TO IMPRISON PEOPLE FROM BROOKLYN IN 2003, FACILITATING A MASS MIGRATION TO PRISONS IN UPSTATE NEW YORK. 95% EVENTUALLY RETURN HOME.

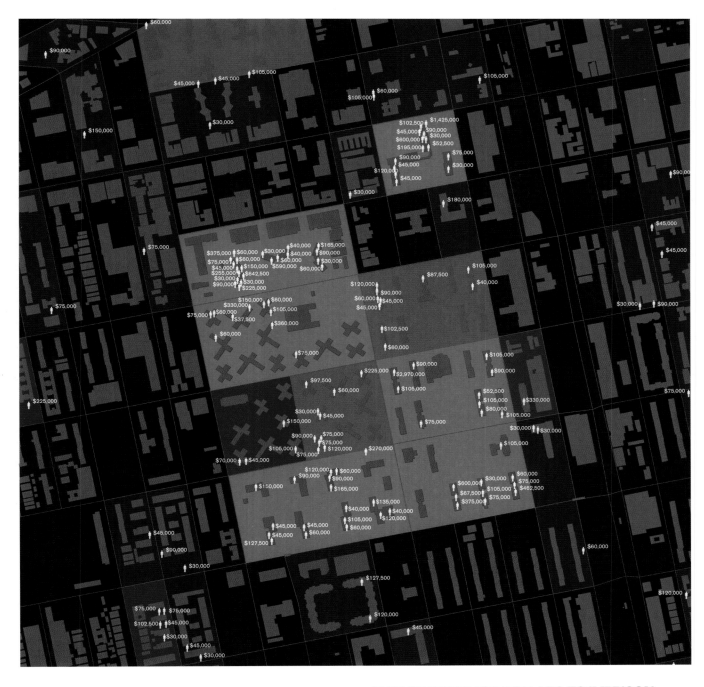

BROWNSVILLE, BROOKLYN

IT COST 17 MILLION DOLLARS **TO IMPRISON 109 PEOPLE FROM THESE 17 BLOCKS IN 2003. WE CALL THESE** MILLION DOLLAR BLOCKS **. ON A FINANCIAL SCALE PRISONS ARE BECOMING THE** PREDOMINANT GOVERNING INSTITUTION **IN THE NEIGHBORHOOD.**

4.13 This map, seen on the wall of the Design and the Elastic Mind Exhibition at MoMA in 2008, shows that more than $17 million was spent on these blocks in Brownsville, NY. Source: Maps created by Eric Cadora, Laura Kurgan, David Reinfurt, and Sarah Williams, courtesy of Spatial Information Design Lab.

The message was delivered using multiple forms of communication. At a 2006 exhibition of the Architecture League in New York City, maps of New York (New York), Wichita (Kansas), New Haven (Connecticut), New Orleans (Louisiana), and Phoenix (Arizona) were presented to the public (figure 4.16). A video seen upon entering the exhibition showed the flow of those incarcerated from their homes in Brooklyn to prisons in upstate New York, which amounted to a mass migration of people. The video asks the public to imagine if just "one million dollars was spent on resettlement rather than imprisonment," what the communities might look like. While the video did not suggest a policy action, it was meant to teach the public about a topic, ask them to rethink it from their own perspective, and ultimately provoke a debate on mass incarceration and its effects on the communities once home to the incarcerated. The Architecture League exhibition helped the project gain recognition of the larger arts and architecture community, and was included in the "Design and the Elastic Mind" exhibition at the Museum of Modern Art in New York in 2008. Now part of MoMA's permanent collection, the maps have been widely exhibited, discussed and disseminated (figures 4.12 and 4.13).

The Million Dollar Blocks visualizations not only supported a public debate through exhibitions, but the work was also published online, in magazines, and other multimedia forums. The simplicity of the maps and image format allowed the message to be easily exchanged through different forms of media. Other groups who sought to make maps of their own city, most notably in Chicago in 2015, imitated the visualization style (figure 4.14).[24] The ability to acquire these images openly online allowed them to be used for all kinds of arguments, including to support congressional funding for the Criminal Justice Reinvestment Act of 2009, which allocated funding to prisoner re-entry programs. The maps brought to light what many criminal justice policy experts already knew, which many in the general public had yet to understand.

The Million Dollar Blocks project could not have been achieved without the collaborative work from our diverse team of criminal justice policy experts, city planners, architects, data scientists, and graphic designers.[25] Acquiring the data came through the relationship that our partners at the Justice Mapping Center had developed during years of work with the Council of State Governments, which works with federal and state corrections departments on a range of policies to reduce costs in state prisons across the

4.14 In 2015, a Chicago organization copied the visual style (red on black) of the Spatial Information Design Lab's 2006 Million Dollar Blocks project. However, not all copies are perfect, and the map has a few flaws, including a missing legend, which makes it hard to know if the data was ethically collected. Source: DataMade, "Chicago's Million Dollar Blocks," Chicago's Million Dollar Blocks, January 7, 2019, https://chicagosmilliondollarblocks.com/.

country. There was much concern about exposing the privacy of the people the data represented, as there should be, and it was our previous work with the Justice Mapping Center that allowed them to trust that we would use the data ethically. Our diverse team, brought together by Laura Kurgan, used its expertise to develop the project, which involved synthesizing data, developing the graphic style, and contextualizing the work with current policies and outcomes of the form of cities. We made our materials available

4.15 Workshops gathered community members, urban planners, criminal justice policy experts, and data scientists to ask how they might use just $1 million to change Brownsville in Brooklyn, where an overwhelming majority of prisoners come from. Source: Maps created by Eric Cadora, Laura Kurgan, David Reinfurt, and Sarah Williams, courtesy of Spatial Information Design Lab.

4.16 Exhibition at the Architecture League. Source: Maps created by Eric Cadora, Laura Kurgan, David Reinfurt, and Sarah Williams, courtesy of Spatial Information Design Lab.

in several formats, helping the work spread further as it tapped into multiple networks and delivering the message of the Million Dollar Blocks project to broad groups of people.

The team engaged in extensive discussion on the ethics of the presentation of the data—which included a discussion of potential biases. The maps were meant to persuade, so we were well aware of our inherent bias. Therefore, we developed numerous versions of the maps making sure everything—from the way we binned the colors in the legend to how we normalized the data—was presented responsibly. We considered it essential to have criminal justice policy experts evaluate the ethics of the maps, as we did not want to further marginalize the populations represented in the maps.

More than a decade has passed since our team developed the maps, and I continue to learn about the varied ways the message of the project impacted others. In July 2019, I sat on a panel with Ifeoma Ebo, a Design Advisor for the Mayor's Office of Criminal Justice in New York City; she spoke about work she is doing to help people returning from prison, and she mentioned Million Dollar Blocks as an influence. Million Dollar Blocks was an inspiration not only to the government of New York City but also to criminal justice advocates as far away as Australia, where the government used the research and maps as evidence for justice reinvestment policies in 2018.[26] The messages of maps do have the ability to travel far.

Interactive Websites Help to Create User-Centered Data Narratives

Much of the criticism revolving around the power dynamics and bias associated with visualizations occurs because it is hard to communicate the complexity of the world through data visualizations that can only provide one vantage point or interpretation. All data has multiple interpretations. Some early innovators in the fields of data for society struggled with this as well. Perhaps the most notable among them was Patrick Geddes (1854–1932), a social statistician of the Scottish tradition and a contemporary of Charles Booth, maker of the famous Booth maps mentioned in chapter 1. Geddes believed surveys were essential to town and regional planning, but he felt limited by the one-dimensional nature of many surveys of his time and encouraged the development of cadastral map sets that included "geography, geology, climate, economic life, and social institutions."[27] Geddes struggled with the limitations of maps as a representational tool; he believed they could not provide the real interplay between the various dynamics that make cities thrive.[28] Contemporary data scientists have more ways to present data: web maps and interactive visualizations, now commonplace, allow us to explore these relationships and find our own meaning in the data. One can imagine how Geddes might have enthusiastically deployed these new digital technologies to illustrate his beliefs on the interface of nature, built form, and its inhabitants.

Geddes's calls, in writings and lectures, for new representational strategies often fell on deaf ears, mostly because his arguments were extremely difficult to follow—and he was said to have a difficult and argumentative personality.[29] Yet, at the heart of his position was the belief that to make planning decisions we must truly understand the people, the geography, and how the two interact, all the while involving the public in the practice of observation. These ideas have had a lasting impact on regional town planning and have influenced many regional planners (including Lewis Mumford) in addition to the methods laid out in these pages.[30]

Geddes is one of the first to bring out the limitations of data analytics, namely the problem of trying to quantify the qualitative nature of cities. While he himself advocated for the development of data, he investigated alternative ways of communicating the insights found within through data visualization. His most notable method of melding the psychological, social, and physical elements that make a city successful is his diagram "The Notation of Life" (figure 4.17), which charts how the relationships between a city's physical form and its citizens' social interactions can result in a flourishing civic life.[31]

Geddes's struggle with the one-dimensional presentation of data provided by maps is similar to that of today's Critical Cartographers, who believe we must interrogate the reasons maps are developed in order to expose the power dynamics that might not otherwise be apparent on a visual reading. Earlier in this chapter I mentioned Jeremy W. Crampton, who argues in his article "Maps as Social Construction: Power, Communication, and Visualization" that one way to account for the power dynamics and inherent bias in the representation of data is to develop ways in which map readers can explore the data and understand it from their own perspective. Engaging data in this way has been made easier with the development of interactive websites that facilitate data exploration. Crampton also believes cartographers should be more sensitive to how the public might interpret the information in maps

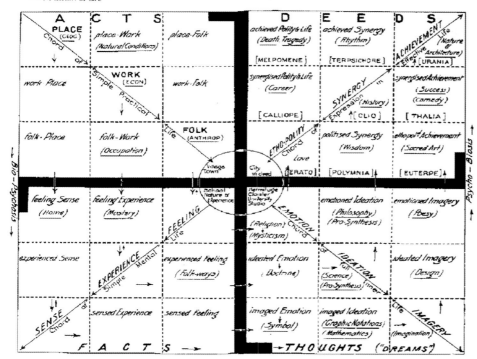

4.17 Patrick Geddes's "The Notation of Life." Source: Geddes, *Cities in Evolution* (London: William and Norgate Limited, 1949).

as misleading or marginalizing. He asks mapmakers to critique their maps through the eyes of the public.

Local Lotto: A Web Tool for Data Literacy

Narratives, images, and interviews are also data, and figuring out how to bring these diverse data sets together is important for generating evidence to help change policy. This is exactly what a group of high school youth did in 2014 when they participated in Local Lotto, a module of a math curriculum called City Digits, to create arguments for whether the lottery provided a benefit to their community. The City Digits curriculum, a collaboration with my research lab (Civic Data Design Lab), CUNY's Brooklyn College, and the Center for Urban Pedagogy (CUP), is developed on a web-based interface that helps high school students build data literacy by allowing them to collect, explore, and form opinions about social justice topics they observe in their neighborhood every day. The Local Lotto module gives students the opportunity to investigate quantitative and qualitative data about the state lottery using interactive maps and participatory data collection. Beyond supporting youth to form opinions about a local and socially relevant topic, Local Lotto's interactive format led to new forms of equitable classroom participation and learning (see figures 4.18–4.24).

Local Lotto comprises four sections that use an interactive web tool. First, students learn how to calculate the probability of winning a jackpot lottery game. Second, they conduct and collect interviews of lottery players and retailers in their neighborhood using the Local Lotto tool on mobile tablets (figure 4.18). Third, students analyze citywide and local level lottery data obtained from the New York State Lottery Commission and public data from the 2010 Census, using an interactive map. Fourth, students synthesize qualitative interview data with quantitative map data to formulate their own opinions about the lottery's social impact. Using the web-tool, students create multimedia narratives called "tours" to teach others about what they learned. Developing the tours allows students to synthesize their data explorations and form opinions, which is an important component of data literacy (figure 4.24).

In Local Lotto, thematic maps highlight the social dynamics of the lottery by visually communicating quantitative data overlaid on a map of the city by neighborhood. Ultimately, the maps show a pattern of increased lottery spending, relative to income, in low-income neighborhoods. Photographs of the local streetscapes and interviews with pedestrians allow for an interpretation of this pattern from a first-person perspective (figures 4.21

4.18 Students who participated in the Local Lotto curriculum conduct interviews in their neighborhood. Source: Laurie Rubel.

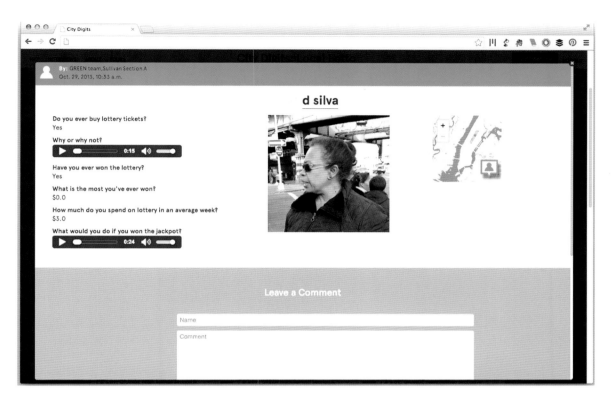

4.19 As part of the Local Lotto curriculum, students interviewed people in their neighborhood who played the lottery. The interviews went directly to the map as a voice recording of answers to questions that the students developed. Source: Screen shot courtesy of Sarah Williams and Laurie Rubel, City Digits, http://citydigits .mit.edu/.

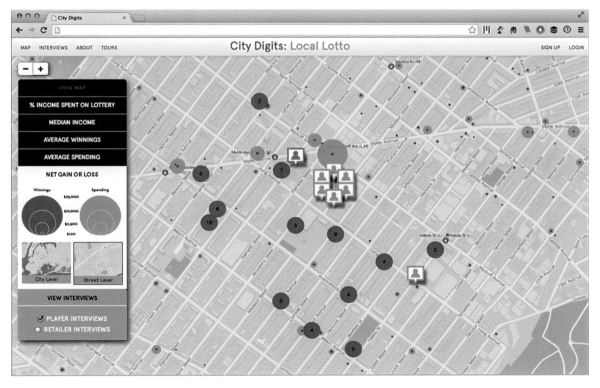

4.20 Interview icons sit on top of a map of lottery data. Source: Screen shot courtesy of Sarah Williams and Laurie Rubel, City Digits, http://citydigits .mit.edu/.

4.21 This Local Lotto map shows the percentage of income spent on lottery tickets. The darker blue represents areas where a greater percentage of median household income is spent on lottery tickets. Source: Screen shot courtesy of Sarah Williams and Laurie Rubel, City Digits, http://citydigits.mit.edu/.

4.22 Net gain or loss map, zoomed into the street level. The green circles represent the amount of money spent on lottery tickets at individual retail stores. The purple circles represent the amount of money won from lottery tickets at the same stores. Source: Screen shot courtesy of Sarah Williams and Laurie Rubel, City Digits, http://citydigits.mit.edu/.

4.23 Students taking part in the Local Lotto curriculum work together to construct arguments using data as evidence. Source: Laurie Rubel, City Digits.

4.24 This Local Lotto student tour explains that lottery marketing represents an "omission of the truth." Source: Screen shot courtesy of Sarah Williams and Laurie Rubel, City Digits, http://citydigits.mit.edu/.

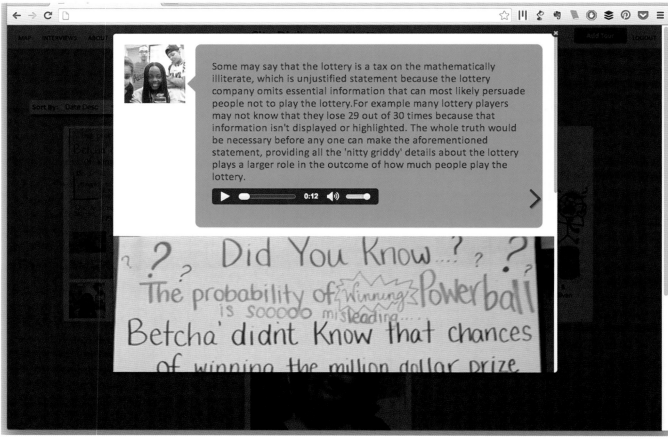

Some may say that the lottery is a tax on the mathematically illiterate, which is unjustified statement because the lottery company omits essential information that can most likely persuade people not to play the lottery. For example many lottery players may not know that they lose 29 out of 30 times because that information isn't displayed or highlighted. The whole truth would be necessary before any one can make the aforementioned statement, providing all the 'nitty griddy' details about the lottery plays a larger role in the outcome of how much people play the lottery.

? ? Did You Know...? ? ?
The probability of Winning Powerball
is sooooo misleading.....
Betcha' didnt Know that chances
of winning the million dollar prize

through 4.24). Combining geographic analysis with images and interviews contextualizes data and allows for the investigation of a civic issue from multiple vantage points.

Local Lotto focuses on teaching data literacy by combining quantitative data (in the form of thematic maps) with qualitative data (in the form of student-collected interviews) and photographs. Blending quantitative and qualitative investigation enables a more nuanced analysis of the lottery, and the use of maps helps reveal relationships between geography and social issues. One student group argued that the lottery was a "scam because it targets low-income areas."[32] Using a map to make the argument they combined the information with interviews they had recorded, in which local deli owners mentioned they thought it was sad that customers with so little money spent it on the lottery.

Inherent in teaching data literacy is the ability to generate arguments about quantitative and qualitative data.[33] In Local Lotto, these arguments tackle whether the lottery is "good" or "bad" for communities. The web tool facilitates the development of argument by allowing students to explore the data on a map. The discussion that ensues while students construct their arguments helps to teach them the policy issues behind the lottery and create a dialogue that is important for community engagement.[34]

The interactive and participatory nature of the curriculum and tool allowed youth to deeply explore data they could connect with. Students indicated that they worked harder in the Local Lotto class than they did in regular mathematics classes because the content was both challenging and relevant to their own lives. Students who were learning English at the same time, and who are typically challenged to participate in class, were able to take leadership roles in the data collection. Some students reflected on the personal significance of their learning during Local Lotto and indicated that they shared their findings at home with family members who regularly buy lottery tickets. Local Lotto taught the youth math as well as how to debate a topic important to society. It also taught students how to take control over narratives around data by allowing them to create their own stories.

Data Stories Create Evidence for Public Debate

Newspapers and other media have long been used as forums for public debate, and many news agencies have begun to use data visualizations as a way to communicate policy issues. Online media's ability to develop interactive "data stories" allows users to explore data, ask their own questions, and generate new conversations. These visualizations do not provide single story lines; rather they allow readers to find their own. The resulting narratives help create a debate on civic topics.[35]

One notable case in which a *New York Times* interactive map generated public debate was an infographic titled "How Minorities Have Fared in States with Affirmative Action Bans," to the point that the graphic appeared in documents used in the Supreme Court. The graphic, developed in April 2014, allowed readers to explore the percentage of Hispanic and black students accepted to state universities in Washington, Florida, Texas, California, and Michigan before and after affirmative action bans were established. At the time the graphic was published, the Supreme Court was set to hear the case of *Schuette v. Coalition to Defend Affirmative Action* (2014), which would decide whether to uphold the affirmative action ban that had been in place in Michigan since 2006. The *Times* generated the visualization so readers could better understand the questions put before the Supreme Court. Overall, the charts and accompanying graphics showed decreases in minority enrollment after affirmative action bans were set in place.[36]

When Supreme Court Justice Sonia Sotomayor used the graphic as part of her official dissent to the ruling to uphold the affirmative action ban, the work became part of a historic debate. She used a screenshot of the *New York Times* visualization to support her position that the removal of affirmative action does, in fact, affect minority enrollment in college (figure 4.25). She stated: "The proportion of black students among those obtaining bachelor's degrees was 4.4 percent, the lowest since 1991."[37] Using the graphs as evidence, she went on to say, "At UCLA, for example, the proportion of Hispanic freshman among those enrolled declined from 23 percent in 1995 to 17 percent in 2011, even though the proportion of Hispanic college-aged persons in California increased from 41 percent to 49 percent during that same period."[38] Sotomayor has always been a supporter of affirmative action, so the graphic most likely did not inform her decision, but she hoped to persuade others by illustrating through data the relationship between policies and minority enrollment.

The *New York Times* received some criticism that these graphics were biased toward lifting affirmative action bans. One article in the *Independent Voter Network* claimed, "All of these numbers have left off something

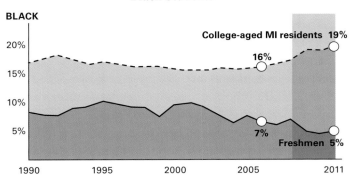

Cite as: 572 U.S.__ (2014)   51

SOTOMAYOR, J., dissenting

UNIVERSITY OF MICHIGAN
Black Students[17]

BLACK

College-aged MI residents 19%

16%

7%

Freshmen 5%

1990   1995   2000   2005   2011

4.25 Screenshot from Justice Sonia Sotomayor's dissent in the Supreme Court case to uphold a Michigan ban on affirmative action. It is a direct copy of a map she saw on the topic in the *New York Times*, showing how newspapers can help to create public debate. Source: Sonia Sotomayor, *Schuette v. Coalition to Defend Affirmative Action*, No. 572 U.S. 2014.

very important; the students that leave those states to attend a university in another state and the students from other states that go to those particular universities. Should they be used to help show if a university is measuring up in minority enrollment to the state's total minority population?"[39] Others found the arguments misleading because they believed the data should have compared college-ready minority students rather than college-age students.[40] Still other articles point out that the *Times* did not include Asians or other minorities in their definition, and, therefore, the story was misleading.[41] Those who supported the Supreme Court's decision often reused the same data visualizations to argue their position. The graphics' appearance in the *New York Times* caused them to be reappropriated and used as evidence for different stories. In short, the graphics generated a debate, which is key to public engagement.

"Graphic storytelling has become an important part of journalism," says Jeremy White, a graphics editor for the *New York Times* who creatively incorporates pictures, video, narratives, graphs, and maps into the stories he develops for the *Times*. The web helps to bring these multiple forms of communication together, allowing users to experience both qualitative and qualitative narratives. The *Guardian*, one of the United Kingdom's main newspapers, does a great job of this kind of storytelling as well, and has developed numerous visualizations that have generated public debate, even here in the United States. One such visualization was "The Counted" (2017), which has collected data on deaths by members of the police force across the United States. *Guardian* readers can slice and dice the data in various ways, finding new insights and reading the narratives from the news stories about the incidents (figure 4.26). What's interesting about "The Counted" is that it did so much more than present the startling story of police killings, but that the data, systematically collected by the *Guardian*, served to change US policies. The *Guardian* mined data from websites, looking for incidents of assaults, and found many more than were reported by federal and state agencies across the country—a marked underreporting. The problem is indeed so widespread that the true cause of death often doesn't make it to the National Vital Statistics System (NVSS). A 2017 Harvard study showed that NVSS was undercounting 55.2 percent of these deaths and that the errors were higher in low-income neighborhoods.[42] The *Guardian* compared its database to those of the Bureau of Justice Statistics (BJS) and even FBI reports, and it reported again on a mismatch. The press about the topic caused all of these agencies to change how they reported data by developing visualizations.[43]

4.26 Screenshot of the "The Counted: Tracking People Killed by Police in the United States." Source: *Guardian*, 2016, https://www.theguardian.com/us-news/series/counted-us-police-killings.

Stop-and-Frisk Brought to an End Using Data

Sharing data has the power to create the public outcry needed to eliminate discriminatory government and law enforcement policies. Under President Obama's open data policies, local and federal governments made available previously inaccessible data, and the public used this data to counter government narratives and change policy.[44] A great example is the New York Police Department's (NYPD) stop-and-frisk data, or Stop, Question, Frisk (SQF), as it was referred to by New York City Mayor Michael Bloomberg, who enacted the policy. This policy allowed police officers to question and search individuals for anything the officers considered to be "suspicious activity"—a category interpreted broadly and with high subjectivity.[45] The policy came into contention when data analysis of UF-250 forms, which were used by the police to report these stops, showed that only 20 percent of stops were warranted. It should be noted that a later review of the data by a federal judge in 2013 found the figure conservative—in reality, most stops were not warranted, showing how the SQF policy was being abused by the NYPD to unduly search members of the public.[46]

Perhaps what is more interesting about the stop-and-frisk story was not the evidence presented at the court case, but rather the fact that evidence was obtainable at all. In 2008 the Center for Constitutional Rights (CCR) performed an analysis of stop-and-frisk data from 2003 to 2007 that was available through new open data policies. They found that stops increased 200 percent in just those four years, and that of those stops, 88 percent did not lead to crimes. They also found that 85 percent of the people stopped were black or Latino, news that gave rise to concerns about racial profiling. In conversations with policing advocacy groups the CCR decided they needed to take action. In 2008 the CCR filed a class action lawsuit, *Floyd, et al. v. New York City, et al.*, arguing that the police were using racial profiling tactics when determining which people on the street to stop.[47]

CCR's mission is to use litigation to fight for social justice, and as they started their work on stop-and-frisk they engaged numerous organizations already working toward police reform. These organizations ultimately recognized their shared interest and created the Communities United for Police Reform (CPR) in 2011 to coordinate as one unified organization to advocate for the process.[48] CPR worked to prove the legitimacy of the case, but also to educate the public using the skill sets of its diverse members, each using their individual advocacy techniques.

The New York Civil Liberties Union (NYCLU) performed additional analysis on 2011 NYPD data, again showing the police were stopping higher proportions of minorities. A report they released in May 2011 found that of the 684,330 people stopped in 2011, only 12 percent were arrested or received summonses.[49] This helped create a counter-narrative to the police department's argument that the policy helped prevent crime. The report, which used visualizations to explain the findings, was widely discussed in the media, including mentions in the *New York Times*, *Forbes*,[50] and the *Wall Street Journal*, among others. These press campaigns helped to get the topic on the agenda of the next mayoral campaign and it was heavily discussed during debates. Bill de Blasio, who eventually became New York City's mayor in 2013, campaigned to do away with the policy.

Beyond the initial visualizations created by the New York Civil Liberties Union, media outlets across New York City used the same stop-and-frisk

4.27 This pie chart from the NYCLU May 2011 report on stop-and-frisk shows that the majority of people stopped were Latino or black.

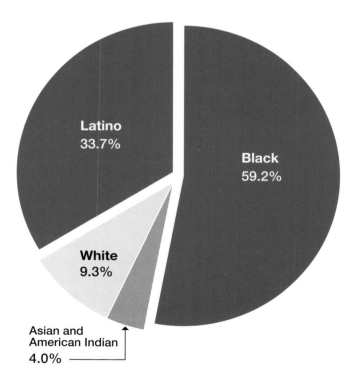

Stop-and-Frisk in 2011, by race

Latino
33.7%

Black
59.2%

White
9.3%

Asian and
American Indian
4.0%

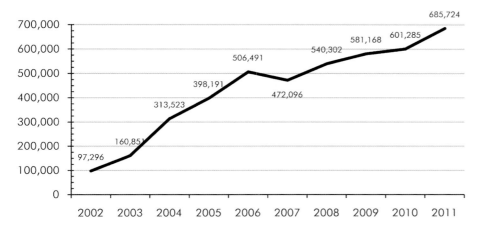

Stop-and-Frisk
Number of Stops Over Time

4.28 This graph from a May 2011 NYCLU report shows the rise of stop-and-frisk practices.

data to help illustrate this unjust police practice. This couldn't have happened were the data not openly available. An independent magazine on Brooklyn issues, *BKLYR*, created a visualization that allowed users to break down the stop-and-frisk data using multiple variables online (figure 4.29).[51] The data visualization was widely circulated among news outlets including a feature in *Wired* magazine; it was retweeted, spreading it across the web. WNYC, New York's flagship public radio station, created the most mentioned visualization, a map (figure 4.30) on which the hot pink blocks represent over 400 stop-and-frisks, or an average of more than one a day. The visualization shows how some minority communities are targets for the policy.[52] It should be noted that the WNYC visualization has also been criticized for its representation style because it has led some map viewers to believe that there are more crimes and guns in these neighborhoods. It was not made clear to the viewers that people in those neighborhoods were stopped more frequently (and often unjustly so), creating a bias in how the data was represented through the map. It is therefore important to be careful in how you construct the narrative around data visualizations. There were also more experimental visualizations, such as the one by artist Catalina Cortazar; she created something akin to a sand

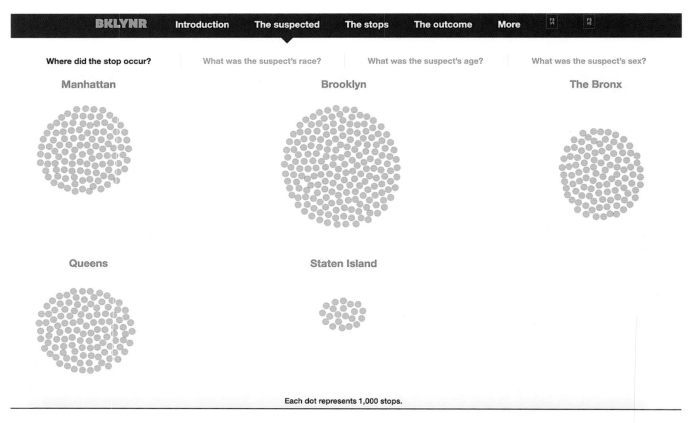

Where did the stop occur?　　What was the suspect's race?　　What was the suspect's age?　　What was the suspect's sex?

Manhattan　　　　　　　　　　Brooklyn　　　　　　　　　　The Bronx

Queens　　　　　　　　　　Staten Island

Each dot represents 1,000 stops.

4.29　BKLYNR blog's stop-and-frisk data visualization, in which each dot represents a proportion of the population so users can easily and visually interact with the data. Source: "All the Stops," BKLYNR, https://www.bklynr.com/all-the-stops/.

clock, and when she turned the sculpture upside down, the sand that couldn't fall through represented the proportion of the population of whites, blacks, and Latinos who were stopped in New York City.[53]

Ultimately CCR filed a federal court case against the policy, and the trial lasted nearly nine weeks. In May 2013 the court ruled the policy unconstitutional under the Fourth and Fourteenth Amendments.[54] "[The City has] received both actual and constructive notice since at least 1999 of widespread Fourth Amendment violations occurring as a result of the NYPD's stop and frisk practices," the decision stated. "Despite this notice,

Zoom to an address

Soundview-Bruckner
Bronx Tract 004800, Block 1002

All stops in this block: **791**

Colors reflect total stops in one block:

1   100   200   300   400   +

Dots mark stops where a gun was found. Bigger dots indicate groups of stops. All data 2011.

Tweet   Like

Full Page | Story | Embed This

Map: John Keefe / WNYC
Data: NYPD Stop, Question & Frisk Data

4.30 WNYC produced a map of the openly available data, which was widely spread across the internet. This map was criticized for trying to make a correlation between Stop-and-Frisk and locations where people had illegally obtained guns. The problem is that Stop-and-Frisk tickets were disproportionately issued in certain neighborhoods specifically because the tactic was overused in minority neighborhoods. Therefore, the data is biased in a way that makes it appear as though minority neighborhoods have more guns. Source: John Keefe, "Stop & Frisk | Guns," WNYC, 2013, https://project.wnyc.org/stop-frisk-guns/.

they deliberately maintained and even escalated policies and practices that predictably resulted in even more widespread Fourth Amendment violations. . . . The NYPD has repeatedly turned a blind eye to clear evidence of unconstitutional stops and frisks." The document further demonstrated that evidence clearly pointed toward systematic racist practices.[55] The continuation of SQF was found illegal and the New York City Police Department stopped the practice. In 2014 the *New York Times* created a feature that showed what removing stop-and-frisk looked like, neighborhood to neighborhood, illustrating a huge reduction in stops (figure 4.31). The policy tarnished the

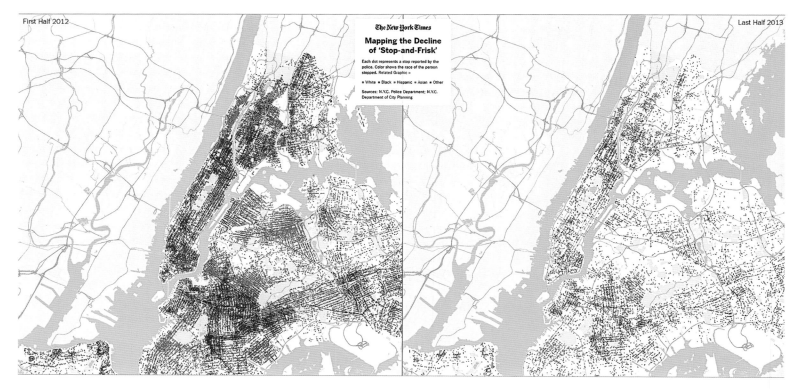

4.31 The *New York Times* data visualization that compared stop-and-frisk after a 2013 federal court case declared the policy unconstitutional. The colors represent the race of people who were stopped (green for whites, blue for blacks, orange for Hispanics, pink for Asians, and brown for others). Source: Mike Bostock and Ford Fessenden, "'Stop-and-Frisk' Is All but Gone from New York," *New York Times*, September 19, 2014, https://www.nytimes.com/interactive/2014/09/19/nyregion/stop-and-frisk-is-all-but-gone-from-new-york.html.

reputation of Mike Bloomberg during the 2020 presidential campaign, as the public found his initial support of the policy hard to reconcile.

The ways in which data was involved in removing New York City's stop-and-frisk policy shows how data developed for power and control can be transformed into something that becomes a tool to overturn unethical practices. It was not merely sharing the data but also the narrative developed around the project that allowed for this reversal. Built collaboratively with policy experts and the community, and delivered to their vast networks, it captured the attention of the mainstream press and media. In turn, the media helped expand the number of people who were aware of the policy and its harms, making it a topic of public debate. Represented in this example are the Data Action methods discussed throughout the book. The stop-and-frisk case shows how anyone can creatively find openly available data, collaborate with experts to tell the data's story, and communicate the result to the public through graphics, ultimately generating societal change.

Sharing Data Creates Transparency, Public Participation, and Collaboration

President Obama once stated that open government seeks "transparency, public participation, and collaboration."[56] Data for policy action needs to abide by those same principles to make ethical uses of data strategies to maximize policy change. Sharing data creates transparency, and this allows those with whom you are sharing data to build trusting relationships that can further mutual policy initiatives. This was true for the Digital Matatus project—the more we shared data with our partners, the more trust we created from and for our mutual work.

Transparency also means exposing data and all the biases within it. This means that through analysis of the data we can illustrate underlying systems of control. Both the prisoner intake data of the Million Dollar Blocks project and the stop-and-frisk story show how you can take a data set used for power and control, and transform it for use by the people whom it represents to exert their own power. In both of these cases the over-representation of communities of color in the data suggests racial bias in policing. In the case of the Million Dollar Blocks, many of the neighborhoods lack the resources to pull themselves out of a cycle of incarceration—jail is the policy answer. In the case of stop-and-frisk, the neighborhoods and the people who live in them are seen as suspicious, which heightens the risk of over-policing and targeting communities of color.

Sharing data through visual interpretation might make underlying policies more transparent, but all data visualizations hold the bias of the creator. This is why public participation is essential for using data visualizations to create policy action. By asking the people who have a stake in the visualization to scrutinize the results, we are made more aware of the potential biases in the data visualization.

Building data visualizations with collaborators helps us identify our unconscious biases, creates trust in the message the visualization is trying to communicate, and, most importantly, disseminates that message to a wide audience, giving our work with data a greater overall impact. Such collaborative effort is the special sauce for most of the projects described in this book. In Nairobi the co-creation of the maps helped to ensure the graphic language was accessible and also built trust among stakeholders, so much so the government trusted the map enough to make it the official map of the city. With Million Dollar Blocks, by collaborating with policy experts we were able to create consumable media messages about the topic. Our collaborative efforts helped tap into a network of multiple advocacy groups, which helped to get the story out to broad networks.

Sharing data also gives it a life beyond the purposes it was originally intended. By putting their data in the public domain, for example, the NYPD allowed it to be used for other initiatives—specifically understanding trends in racial profiling. This is true in Nairobi, too, where freely sharing the data meant the local technology community could develop its own technologies with the data. When the New York State Lottery commission shared data with us to teach youth about the policies regarding the lottery, they transformed something used for management into a learning tool. By developing a data set and visualization of those who died at the hands of police in "The Counted," the *Guardian* created an open data set that was later used to show how governmental data was not tracking this problem properly.

Debate is important to our society. Open data is clearly political, and so too are data visualizations. Not only do they help to communicate insights found in data, but the very act of making them also belies an agenda. Whether it is to illustrate and persuade, as in Justice Sotomayor's dissent, or to argue for or against climate change, data visualizations set apart a topic or an idea as something of significance or importance. Data visualizations work so well at communicating policy because they embody a sense of legitimacy, which can sway the public. This is of course a double-edged sword as ultimately the messages data visualizations communicate come from those who design them. Being transparent, involving the public, and collaborating in the creation of data visualizations helps, but as a society we need to become more data literate so that we can critique the messages data visualizations provide.

DATA AS A PUBLIC GOOD

In 2019, I gave a talk at the 10th World Cities Summit (WCS) Mayors Forum held in Medellín, Colombia, about the need to find ways to build data to make planning decisions for many rapidly developing cities in the Global South. A corporate participant in the conference responded, "The problem is not the need of data; we are swimming in data. It is finding the resources to analyze the data we do have to generate policies for cities." This statement represents one of the biggest issues revolving around the hype of big data analytics. Sure, there is a lot of data, but there is a growing gap between those who can access that data and those who can't. One of the reasons this gap is growing has to do with the fact that private companies, for the first time, own and control the majority of data and can decide who to share it with and for what reason. Some of the biggest cities in the world are, in fact, swimming in data, and that's because they have the money to pay for it or collect it for themselves. But for some communities, notably rural and developing cities, creating or buying data is either too costly or there is no political will to collect it. Private companies do sometimes share their data for the betterment of society, but usually only when it benefits them to do so. In this way, they still retain power over the data and its potential message, and the outcome is not necessarily in the best interest of society.

Given the overwhelming amount of privately owned data, governments have started to set standards for ethical data practices—notably, for example, the EU's General Data Protection Regulation—but overall they still seem overwhelmed and unprepared for the data deluge to come. To use data owned by private companies to help society, we must create regulations that protect the people in the data and incentives for data provision, and thus a path for using data for the public good. Government regulations alone cannot be the solution to ethical practices for using data because not all governments use data ethically. In totalitarian governments, for example, data is often used to oppress the population. Regulations also often lag far behind innovation. So, while we should call for new regulations, it is important for those of us working with data to create our own standards for using data ethically. When using data for action, we must explore the ethics around using data for a public good.

Data Scarcity in the Era of Big Data

With all the hype around the use of big data, especially the data collected and owned by private corporations, it's often hard to imagine that in some parts of the world even basic demographic data is extremely difficult for governments to obtain; and in many of those areas data is only available if private corporations choose to provide it. According to an article in the *Economist* in 2014, "Africa is the continent of missing data. Fewer than half of births are recorded; some countries have not taken a census in several decades."[1] The World Bank confirms this assessment in a recent report, titled *Data Deprivation: Another Deprivation to End*, explaining how data is so scarce in developing countries that it makes it difficult to track simple, sustainable development goals, including poverty reduction.[2] According to this report, fifty-seven countries have only one or no poverty estimate; this amounts to roughly a third of the world's developing or middle-income countries missing essential data to track poverty.[3] One of the main reasons these countries lack such data is that their governments do not have the technological or financial means to collect it. The United Nations report, titled *A World That Counts* (2014), argues that a lack of data can lead to a "denial of basic rights."[4] Indeed, the development of data, then, is essential not only because everyone must be counted but also because data is necessary for governments and NGOs to plan for and improve both infrastructures and social services. The fact is that the needs of those not counted are often forgotten.[5]

This is especially frustrating because that data exists, but not in the hands of people who can use it to track poverty or build urban infrastructure. Private companies, including telecoms, Facebook, and Google, for example, can infer population data by analyzing their data exhaust, but they do not share it with people who could use it to improve society. For private companies to analyze data for such a purpose can be costly, and it is usually only done when there is a mutual benefit; for example, a widespread public health concern. We saw this happen in 2020 as the novel coronavirus pandemic became a global emergency. There are many cases in which NGOs and multilaterals obtain data from private and typically foreign companies and conglomerates to estimate basic social services because governments do not have the information.

Data Is a Public Infrastructure

Now let's look at the same idea in American and European contexts, where private companies currently hold and are set to hold exponentially more data about cities, their populations, and their communities. Let's use Waymo, Google's autonomous vehicle program, as an example. A single Waymo test vehicle scans the environment with LIDAR sensors producing about 30 terabytes of data per day—that's three thousand times the amount of data that Twitter produces daily.[6] Programmers at Google use this data to construct 3D representations of the physical world, often referred to as digital twins, which are used to guide autonomous vehicles on the road. In this virtual environment, autonomous vehicles—or any robot for that matter—can be guided and instructed to turn, to stop and pick up a passenger, or to come to a halt for a pedestrian to cross the street. Now imagine that this data is combined with all the data Google stores about you—what purchases you have made online, where you work, the last concert you went to, the locations of your upcoming vacation, even your political views. When this data is added to the digital world created for the car, it can be used to personalize our driving experience. Having captured your buying habits and movements, this data can be mined to identify, for instance, when and where you might want to buy a quart of milk, and your car can be programmed to remind you to do it. Adding this type of information to autonomous vehicles' databases allows cars to make decisions based on previous human behaviors.[7]

This alternative reality—really a new digital reality—will be the infrastructure of the future, itself a new "public good." We will tap into it to perform all kinds of tasks. Much in the same way networked computers created the World Wide Web, this environment will power many innovations created by those who have the capabilities to tap into it. Right now, that ability lies solely with the companies who generate the data, companies such as Waymo, Tesla, Ford, and others experimenting with autonomous vehicles. Companies are already using their virtual worlds to create diverse products beyond the car. Google has developed an augmented reality navigation tool for your smartphone, fueled by this digital environment (figure 5.1a and 5.1b). Microsoft has started to use some of this data to create a product it calls 3D Soundscapes (figure 5.2), which taps into the virtual environment

5.1a and 5.1b Google's AR navigation application adds a digital fox (left) to your navigation space and suggests places to eat (right). Source: James Kanter, "Google Showcases Augmented-Reality Navigation on Google Maps," *Business Insider*, May 9, 2018, https://www.businessinsider.com/ google-showcases-augmented-reality -navigation-on-google-maps-2018-5.

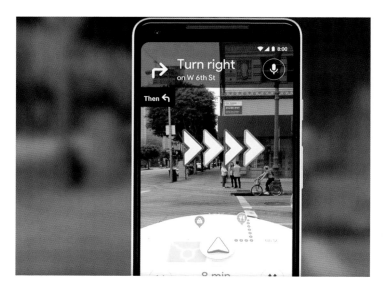

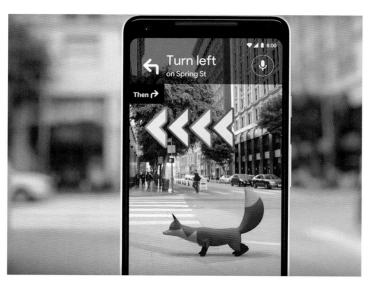

5.2 3D Soundscape created by Microsoft allows blind people to navigate the city. Source: Adam Peters, "This Device Creates A 3-D Soundscape To Help Blind People Navigate Through Cities," *Fast Company*, November 20, 2014, https:// www.fastcompany.com/3038691/this -device-creates-a-3d-soundscape-to-help -blind-people-navigate-through-cities.

to help a person with vision impairment or loss to navigate through the city. The applications for this new virtual environment are almost endless.

Whether it is in Africa or North America, data runs the world we live in and is essential for contemporary life. I would argue that certain data should be considered a public good, and we must set some ground rules for how to access and use it for the betterment of society—ethically and responsibly.

Data Colonialism

Where once governments were the keepers of data, now private institutions are in control. Scholars have called the shifting relationship in who owns our data a new form of imperialism, and refer to the asymmetry in data ownership and the power that comes with it as "data colonialism." In a recent article, Jim Thatcher and his colleagues recently introduced the idea as an analogy for understanding "the shifting terrain of data's role in society," as it attempts to highlight "power asymmetries inherent in contemporary forms of data commodification."[8] They point out that when nongovernmental organizations (private and nonprofit) seek to obtain control over infrastructure or other development rights, they now have a bargaining chip, namely their data. In some cases private companies don't even need this bargaining chip as they simply use their data to build their own infrastructure.

Data colonialism doesn't just apply to the relationship between private companies and the developing world, but is also relevant in North American and European contexts as well. In a recent book about the topic, *The Costs of Connection: How Data is Colonizing Human Life Appropriating It for Capitalism*, the authors Nick Couldry and Ulises Mejias define data colonialism as "the extension of the global process of *extraction* that started under colonialism and continued through industrial capitalism, culminating in today's new form: instead of natural resources and labor, what is now being appropriated is human life through the conversation into data."[9] It is we, the world's citizens, who collectively make the data they are referring to by providing large corporations and many smaller ones with detailed information of every moment of our lives: everything from our real-time geo-location to our email records, most of it time-stamped. That data belongs to companies who log and package it to sell it as a product to others for everything from ad targeting to improving transport services. We might have heard it said that "data is the new oil"—and, according to the authors of *The Costs of Connection*, we are little drops of oil, being exploited for profit and control.

Similarly, as the authors of *Compromised Data* express it in their book on the topic: "data is being employed to accelerate prevalent neoliberal redefinitions of the role of the state and the transformation of citizenship into consumer practices."[10] The concept of data colonialism is hard to wrap your head around, so I will attempt to share my view on its meaning. I believe the scholars' point is: because data and many of the insights it offers are gathered by private institutions, responsibilities once attributed to the state are now taken on by private agencies. And, for some governments, the shift of responsibility to private companies is exactly the point: it helps to promote their neoliberal agenda. One example often used as evidence of this phenomenon is the *open data movement*, which was promoted by governments to help spur private innovation for the development of services that governments would typically provide.[11]

These type of concerns about the shifting role of technology in society are not new. In the 1980s, for example, many people were worried about the control mass media and TV broadcasting exerted over shaping society's ideologies. The term *electronic colonialism*—coined by Tom McPhail, a Canadian media scholar trained by Marshall McLuhan—addresses how media companies have sought to capture and control the "psychological empire" of our minds with their input, rather than just becoming empires that literally extract our data.[12] I believe we are still very worried about both types of control, electronic colonization and data colonialism. Our current conversations about the ability of social media to deliver "fake" news and influence our ideas are a perfect example, illustrated by the Cambridge Analytica scandal mentioned in chapter 3. Data has always been a tool of power and control. Perhaps the problem now exists not in the governments that try to control us but rather in private companies we never elected to represent us. To be more precise, governments seem to be systematically giving up control of services traditionally in their hands in favor of private entities—the essence of neoliberalism.

I would argue that data colonialism is not new—we just have new colonizers. Whether it is the state, the landowner, or the private institution, whoever owns the data also retains power and control. The framing of data colonialism, however, helps us understand the current shift toward private entities' control of data and their use of it in extractive ways. Potentially more problematic is that our government may have unintentionally orchestrated this colonization, given that many of the companies whose algorithms now mine our data were created with government investment. Google developed algorithms with funding from the US National Science Foundation. And before that, similarly, the Defense Advanced Research Products Agency (DARPA), an arm of the US Department of Defense, created the internet.

Private Data Serving the Public Good

Is the data owned by private companies a public good and therefore something that should be protected by governments? The notion of private data being a public good has been widely discussed, particularly in the humanitarian sector, where data is scarce and data collected by companies is the only form of information about the population. Many cities benefit from privately owned data with examples from crisis management, public health, transportation planning, and urban and economic development.[13]

Let's pause a moment to clarify the terms in these conversations. Public goods are commodities or services provided, without profit, for the benefit of society (that is, for the greater good). Economists define public goods as non-excludable and non-rivalrous.[14] *Non-excludable* means that rejecting or barring someone from access to the good creates broad, societal costs. This also applies to data: our society cannot function without it. *Non-rivalrous* means the goods don't dwindle as they are consumed, and the cost of producing the good is marginal; this holds true for data in that data-driven companies, with almost every interaction, create some form of digital exhaust. Public goods don't have to be provided by governments, although governments protect public goods *for* their citizens. One example of such a public good is electricity, which is controlled by private companies but highly regulated by government in order to ensure equitable access to it. Economists assert that the costs of public goods should represent the marginal benefit to consumers equal to the marginal cost of producing it.[15]

But economists continue to debate whether data ought to be considered a public good. Both Joseph E. Stiglitz and Hal R. Varian laid the groundwork for such discussions in their individual scholarship, published in 1999, on information and knowledge.[16] They argued that information and knowledge have a low cost of production, and there is no extra cost for others to use them: that makes them non-excludable and non-rivalrous, and therefore a public good. Information and knowledge, however, are not data. Economists are not as clear on whether data is a public good, largely and precisely because private companies control data and can therefore limit access to it, which makes data an excludable resource.[17] For sure, the General Data Protection Regulation (GDPR) established by the European Union makes the data we contribute to private companies *our individual property*, which confuses the issue even more.[18] But can data be a public good if we own it as individuals?

The argument that such data is a public good is further complicated by the fact that private companies have set up entire business models to extract value from data. A great example of this is Uber, the ridesharing company. Uber has released aggregate data on traffic patterns in cities, but that data is not much different from information some cities already collect. For most cities the real value would come from seeing individual trips, which would help them analyze and find ways to alleviate congestion. It should be noted that recently Uber has worked with cities to try to identify ways to provide useful data. Earlier, Uber claimed that giving away its data is equivalent to giving away its business. This is true to the extent that the data's value comes from using that data to predict travel time and cost; sharing it would amount to releasing its secret sauce because others could then perform the same analytics. Not to mention that sharing this data in an unaggregated way brings up a host of privacy concerns. Can Uber ensure that governments will retain the privacy of their citizens? At the same time, Uber is operating its business on public infrastructure, namely public streets and roads. Does it owe the public anything for operating on public infrastructure? What, if any, is its responsibility to the public? Uber, for its part, has felt pressure and is trying to identify ways to share this data.

It's not surprising that companies don't want to give their data away: they spend time and money filling their databases with information, and they have legal precedents to back up their claims. In *Feist Publications v. Rural Telephone Service Co.* (1991), the Supreme Court specifies that a compilation of facts (such as an alphabetical list of names and numbers in the white pages of a telephone directory) are not copyrightable, but creatively reorganized telephone numbers (such as those in yellow-page listings) are the property of the company that spent time and money to *organize* the numbers (into categories such as plumbers, cleaning services, medical facilities, and restaurants).

The legal precedent for the ownership and use of data is startlingly sparse. We must develop stronger regulations regarding how data can be used and by whom, and in what context. Let's take the example of Waymos, the autonomous vehicles I mentioned at the beginning of the chapter. The digital infrastructure that Google creates to run these cars will be the very data that makes autonomous, robotic pickup and delivery possible for *all of us*, regardless of whether we use those services immediately or in the future. If portions of the population are excluded from Google's databases, they will not be part of the service domain—thereby harming those neighborhoods and the fabric of the society to which these neighborhoods belong. Now, combine the fact that Waymo's cars are operating on the road—public property or a shared public good. They are extracting value from operating this way. What, we must ask, is Waymo's obligation to the public or the government that represents that public? Why not share the very data the company is able to extract in return for operating on public thoroughfares?

Cities still have the upper hand to negotiate better deals for working with private companies. Moreover, it is imperative that city leaders use their authority: many cities derive 50 percent of their budget from mobility-related fees—everything from parking violations to bridge tolls.[19] Cities should think of creative ways to regulate autonomous vehicles (AVs) to make sure that their introduction and establishment in the transport infrastructure won't marginalize some populations. The potential of AVs to marginalize stems from their potential to destabilize public transit; if transit is destabilized and potentially collapses, those who can't afford AVs will be left behind, lacking basic mobility to commute to work, take their kids to school, and access city services. We must think of and address the potential risks that innovation can pose to those on the margins of society. Regulations also have to be considered to recover the cost of using the city's infrastructure, such as charging fees for drop-off and pickup zones, developing new taxes for empty seats in vehicles, and fees for parking empty fleets. Some cities have begun to question whether governments or private companies should run autonomous buses. The public must put guidelines in place for civic infrastructure now, before we relinquish all control of data to private companies.

In 2019, I spoke with the Mayor of Kampala, Uganda, on a panel at Transforming Transit, a conference put together by the World Bank and the World Resources Institute. He was set to discuss Uber's role in the city's transport landscape. He spent the majority of time lamenting the operating agreement he had set up with Uber. When Uber came to Kampala the city didn't create any kind of regulation or service fee; they wanted Uber's innovation and didn't want to create any barrier. What he didn't consider was that Kampala's taxis have a high tax rate, close to 50 percent of their overall profit. It's therefore not surprising that taxi drivers began to shift to driving for Uber to avoid the taxes, creating a significant fiscal loss to the city. The mayor reminded other mayors in the room that they need to negotiate more effectively when private companies ask for an operating agreement; that is the only time they have power or control over the situation. With only two years left on the operating agreement in Kampala, the transit landscape has significantly changed, and the city will renegotiate the terms. But the public has gotten used to Uber, making the negotiations that much more complicated.

Privacy for My Business and My Customer

Michael Bloomberg is a huge proponent of using data for the betterment of society, and much of his foundation's work seeks to use data in this way. The Head of Data Science for the Bloomberg Foundation, Gideon Mann, said he believed that private data could have huge benefits for society, for health, development, and environment practices. In 2016, he wrote an article in which he argued in favor of using the vast data sets available to private companies:

> The data being collected and generated at private companies could enable amazing discoveries and research, but is impossible for academics to access. The lack of access to private data from companies actually has much more significant effects than inhibiting research. In particular, the consumer level data, collected by social networks and internet companies, could do much more than ad targeting.[20]

Yet Gideon himself admits that, while this is Bloomberg's position, it is often hard to make happen because private companies need to devote time to making sure the data is secure. Here security means not just protecting the personal privacy of people who might be in the data set, but also valuable business information. Currently there are a lot of costs associated with maintaining such complex security—including legal costs as well as staff time to synthesize the data—and there are no incentives for private companies to bear those costs. So, while companies may have good intentions for providing data as a public benefit, it is doubtful anything will happen unless there are incentives.

Gideon seems affected by what he refers to as the America Online (AOL) "disaster." In 2006, AOL intentionally released for research purposes the log of twenty million searches of around 650,000 users over a three-month period. The problem with the release was that even though AOL subscribers had unique IDs meant to protect their privacy, the freely released data included personally identifiable information that could expose them when the data was cross-referenced with other data sets. In fact, the *New York Times* interviewed one of those AOL users, Thelma Arnold, and with her permission published an article demonstrating how easily the privacy of individual users could be violated and the data trial from their online searches traced back to them.[21] After three days AOL removed the data from its website, but that was a bit late since the data had already made its way around the internet, with copies distributed, reposted, and downloaded by the thousands. AOL's CTO Maureen Govern resigned as a result of the data breach, and a class action lawsuit against AOL was filed in September 2006. The incident triggered debates about data privacy and research ethics.[22] Many argue that it's close to impossible to find fully anonymized data, and that users can nearly always be identified. It is therefore not surprising that it is difficult to find companies willing to release their data, even for the public good.

The issues of personal privacy are often overlooked in humanitarian work, where NGOs seek data from private companies to address critical public health concerns and ask it to be shared in ways that don't protect the privacy of individuals, all in the name of the great public good. Sean McDonald uses the 2014 Ebola crisis as an example, when the World Health Organization (WHO) demanded that call detail record (CDR) data be provided to trace the spread of the disease, without the consent of the people in the data. He argues that this type of data is perhaps the most sensitive information that can be given, and when used in a disaster event the proper protections on the use of this data are not ensured, thereby putting millions of people at risk.[23] McDonald is not arguing against using this data, but rather he is attempting to expose how the humanitarian sector often acts cavalierly in its approach to using highly sensitive data in the name of the public good. He uses this example to encourage us to consider more ethical ways of managing this data during a disaster response. He provides a few recommendations: create transparency in the models being used, thereby allowing the public to engage in the data analysis; create data-sharing templates to ensure the ethical use of the data; create a way for the people in the data to "opt in" (or donate their data) and provide explicit consent. Data Action asks us to use data in this way too: engaging the people in the analysis is essential not only to be more ethical but also to make sure the narrative around the data is presented properly.

Data privacy is an important element of the General Data Protection Regulation (GDPR), which became effective as law on May 25, 2018. This regulation seeks to give citizens and residents of the European Union more control over their personal data. It is meant to maintain stronger protections than the previous Data Protections Direction, established in 1995. Both policies were developed from a set of principles called the "Fair Information Practices" developed by the US Federal Trade commission in 1973.[24] These early ethical frameworks came with few enforcement capabilities. The GDPR comes with an enforceable framework, which protects citizens' control over their privately identifiable information.

Thanks to the GDPR, citizens can request information about the processing of personal data; get access to their personal data; have the data corrected; request that personal data be erased; object to the processing of their personal data; and request placing restrictions on the processing of personal data.[25] Although only enforceable in the European Union, there is hope that companies operating both in the United States and Europe (and elsewhere, too) will adopt privacy rules. In a recent informal test, however, the *New York Times* found that two reporters, one from the EU and one from the US, both of whom asked for their personal data, were given very different levels of information. Both reporters found that not all their personal data was given to them when requested. While the EU reporter received more data, she knew there were still many pieces of information she had not received.[26] Companies may need time to adjust to the regulations because they need the staff to address individual requests, but it still seems arduous to control data housed with private companies even though we may have every right to it.

Given that regulations are still emerging and will always fall behind practice, organizations have emerged to help people use data ethically for humanitarian purposes. One such organization, Responsible Data, was founded by a diverse group of nonprofits including Amnesty International, Oxfam, Aspiration Tech, School of Data, Huridocs, greenhost, Ushahidi, and Universiteit Leiden. According to its website, "Responsible Data (RD) is a concept to outline our collective duty to prioritize and respond to the ethical, legal, social and privacy-related challenges that come from using data in new and different ways in advocacy and social change." I find their work particularly useful because instead of criticizing the potential for people to do harm with data, they provide tools and resources that help us think about how to use data ethically and to help generate ethical work. The guidance is useful. They have developed seven ideas for considering how to use data responsibly, which I believe echo the Data Action methodology. The first, power dynamics, asks us to consider the least powerful actors in data analytics and determine how they might respond to the work we are doing. They ask us to build in ethical checks during the process of taking data to action. They advocate what they call the *precautionary principle*: if you cannot determine the risk or harms, then evaluate whether the work is truly necessary. They believe we should all innovate with data thoughtfully, not for speed. We should hold ourselves to a higher standard, as regulations have yet to catch up with practice; we should develop our own best practices and lead by example—address diversity and bias and ask ourselves what perspectives are missing. Finally, build better behaviors by acknowledging that each situation is different, and rules that work in one area don't necessarily work in others.[27]

Sharing Data When the Benefits Are Mutual

When we investigate the reason companies share data, we are bound to find hidden agendas. This is not always problematic, because corporate agendas can sometimes work toward the betterment of society. But it is essential we be aware of the *incentives* private companies have when sharing their data, so as not to contribute to potential exploitation from using it. The best example of the way in which private companies share data for mutual benefits is the work they do to gather population data. Knowing where people live is essential for making decisions about everything from the development of infrastructure to the provision of social services. The PopGrid Data Collective was organized in 2018 not only to facilitate collaboration among population data producers, whether private or nonprofit, but also to inform potential users that the data exists, to work with users and donors to identify and prioritize data needs, and to promote collaboration on data validation, documentation, and access initiatives.[28] PopGrid's members use a number of methodologies to generate their data, including mobile phone traces, household surveys, and satellite data interpretation.

Perhaps one of the biggest private contributors to PopGrid is Facebook's Connectivity Lab, which has teamed up with Columbia University's Center for International Earth Science Information Network (CIESIN) to develop high-resolution population data. The work, which started in Ghana, Haiti, Malawi, South Africa, and Sri Lanka, has now been undertaken in over twenty-three developing countries. The process involves mapping human settlements using machine-learning techniques to detect buildings in the satellite images (figure 5.3); taking census population data, which usually exists at a regional level; and splitting up the regional population counts among the settlements based on their overall size (measured by number of houses). If one region, for example, has two rural settlements of roughly the same size, and the county in which these settlements exist has enumerated one hundred people, PopGrid estimates that fifty people live in each settlement. To perform this work across whole countries a one-hundred-meter fishnet grid is laid over the whole geographic areas (figure 5.4). If a settlement's location intersects with a grid cell, the cell is given a proportion of the regions overall population counts.[29] Before this, census maps presented data for large regions, but this work now creates population maps illustrating where people specifically live among the vast regions of unpopulated land in between the settlements.

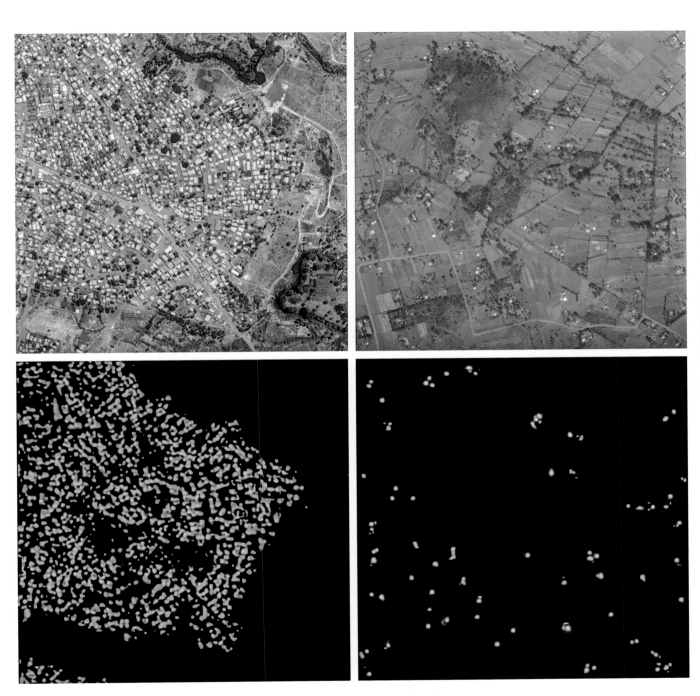

5.3 The top images show satellite data, and the bottom images show where structures have been identified through machine learning techniques. Source: Courtesy of CIESEN, https://code.fb.com/core-data/open-population-datasets-and-open-challenges/.

5.4 Image of Facebook settlement data in Malawi. Population is only attributed to the grid cells where there are structures. Source: Data Blog: World Bank (blog), November 18, 2016, https://blogs.worldbank.org/opendata/first-look-facebook-s-high-resolution-population-maps.

Having data about exactly where people live allows Facebook to locate rural settlements, their populations, and plan for potential services. One can use such data to determine, for example, which villages have access to roads, health care, or internet by performing simple map overlays with other existing data. By identifying the location of settlements using the satellite data and allocating the population numbers based on the size of the settlement, the Facebook/CIESIN team has created a more accurate representation of populations across many rural areas.

Having detailed population distribution data is extremely helpful during disasters. According to Facebook's website, the "Red Cross is already planning to use the population maps to help combat malaria in Malawi, and to target humanitarian aid after natural disasters."[30] The data set is also useful to detect urban growth and was used in Jakarta to define the extent of growth of this rapidly developing region.[31] Again, however, it should be noted that giving away this data is not necessarily an altruist act, as providing data freely will help Facebook develop its business there. Increased development of internet services will ultimately help Facebook's bottom line: as rural areas benefit from improved internet access, Facebook will generate new users and increased user activity.

Facebook's human settlement data has received a lot of interest from researchers. Studies it spawned include understanding water vulnerability in the area known as High Mountain Asia; planning electrification, geomarketing in telecom, and other planning initiatives in Ethiopia; planning electrification in Uganda; and, most recently, the COVID-19 response in Sierra Leone.[32] There are nineteen countries for which demographic data is currently available for download. According to CIESIN people have downloaded data about approximately half of them at a rate of 100 to 300 times per month per country, and about the other half roughly 50 to 100 times per month per country.[33] The countries whose data has been most often downloaded are the five from the initial release: Haiti, South Africa, Sri Lanka, Ghana, and Malawi. The data of Haiti and South Africa have been downloaded more than two thousand times. Detailed, up-to-date data is extremely difficult to obtain for rural areas across the world—which means releasing this data has a great deal of impact.[34]

Developing Data-Sharing Agreements

Population data is essential for planning. So, too, is transportation data: one of the most serious problems in many rapidly developing cities is congestion. The World Bank has tried to overcome data-sharing limitations by creating the Open Transport Partnership (OTP) in 2014. This partnership was established to make it easier for private companies to contribute their data by removing some of the barriers of data sharing. These include the cost to aggregate and anonymize the data as well as to draw up the lengthy legal contracts needed to protect the two organizations regarding the use of the data. The project was started by developing formal relationships between the World Bank and Easy Taxi, Grab, and Le.Taxi, three of the top ridesharing companies in the world, which combined, cover more than thirty countries and have millions of customers.

The World Bank team in collaboration with Grab, the leading taxi-hailing app company in Southeast Asia, developed a pilot open-source platform for using anonymized GPS data generated by more than 500,000 Grab taxi drivers. The platform allowed the World Bank to analyze traffic congestion peak patterns and travel times, with an ease and level of detail that was previously inconceivable. Of huge benefit to Grab was that World Bank data scientists did almost all the work of logging in to Grab's database to aggregate the data in a form that made it redistributable. Performing the data filtering and aggregation so that Grab would not have to bear the cost of doing that work made this a mutually beneficial relationship. According to Grab, not having to pay for the data aggregation had a big influence on starting the relationship and giving away the data.[35]

An example of how the World Bank–created Grab platform was used can be found in the Philippines: with this data the city governments of Manila could, for the first time, answer fundamental questions needed to address safety and congestion: "Where and when is congestion most acute? Where and when are our citizens most vulnerable to road incidents? And, most fundamentally, when we invested in interventions to mitigate accidents or congestion, did these investments pay off? What was their impact? Could we have done better?"[36] The project has been described as "leapfrogging," meaning that the Philippines government did not have to create its own data-monitoring systems but instead relied on the networks of private entities to get the data it needed for transport planning.

5.5 These maps show the density of GPS data available from Grab Taxi. Source: Holly Krambeck, "The Open Transport Partnership" (January 20, 2017), https://www.slideshare.net/EMBARQNetwork/the-open-transport-partnership.

Monitoring transportation costs money, as do the ramifications of congestion. According to one source:

> Congestion in Manila costs the economy more than $60 million per day, and it is not atypical to spend more than 2 hours to travel 8 km during the evening commute there. But beyond these statistics, until recently, very little was actually known about Manila's congestion, because the equipment and manpower required to collect traffic data have far exceeded the available resources. Most cities in developing countries face similar issues.[37]

The example of Manila demonstrates how data collected by private transit companies is essential for analyzing traffic conditions in countries that do not have the ability to create their own data. Ridesharing companies, as in the Grab example, also have an incentive to share it with governments as they are operating vehicles on government roadways. Improving traffic conditions benefits their businesses. Mutually beneficial relationships such as this one between Grab and the World Bank can make data sharing a reality.

Negotiating the Terms of Data Access

While activist advocates of data ethics (and often companies themselves) have called for the creation of common data license agreements to navigate the ethical terrain around data, in reality these agreements have had mixed success. These boilerplate agreements are meant to alleviate the often cumbersome process of developing agreements with private companies, yet they only seem to work well when applied to relationships with large, well-known organizations such as the World Bank or academic institutions. But once these institutions get their hands on the data, they often find that they needed data that was not included in their data sharing agreement—and a new agreement needs to be forged, which takes time.

The World Bank has negotiated license agreements with a range of companies including LinkedIn, Grab, and Waze, among several others. If researchers at the World Bank want to work with Waze data, for example, they must submit a description of their research intentions to the World Bank data team—a request that must include information on the data points they are interested in, what they intend to use the data for, and what time period they are requesting. The World Bank's data science team then approves or turns down their request without having to renegotiate with Waze, which saves administrative time for all parties.

The World Bank is interested in furthering the ability to work with cell phone call detail records (CDRs) and recently announced a partnership with Global System for Mobile Communications (GSMA) at the Mobile World Congress 2018 in Barcelona. The GSMA represents the interests of nearly eight hundred operators with more than three hundred mobile operators worldwide. By partnering with them, the World Bank hopes it can decrease its negotiations with individual operators, thereby also streamlining the process of data sharing.[38] Both organizations are hopeful that their collaboration will "unlock new insights from anonymized data collected by mobile network operators through IoT [Internet of Things] devices and aggregate data from smartphone use. It will also call on industry leaders, development partners and governments to work together in building a strong enabling environment for the IoT while protecting personal privacy."[39]

While the World Bank has been relatively successful at developing pre-negotiated license agreements that benefit their researchers, others have struggled to establish agreements for the data they need. The nonprofit DataKind, which works with corporations and nonprofits to use data analytics for social change, often has a hard time completing two-year contracts with organizations, as it can take close to a year to negotiate the terms of the data licenses. According to its founder Jake Porway, even when a private company contracts him to do research with DataKind, the licensing agreements can take so long to work their way through the company's legal machine that DataKind needs to find other data in order to perform the analytics to meet their contract demands. Natalia Adler, who works as UNICEF and was a Data for Good Bloomberg Fellow, echoed these points. She once believed that pre-negotiated licenses would help her with her work and even sought out a lawyer to help her draw them up, but she found that it was challenging to create a single agreement that most companies would accept. She ended up having to negotiate individual agreements each time she needed a new data set; she was never able to make a boilerplate agreement work.

One way private companies get around the cumbersome data licensing process is by making researchers company employees. This commits the researchers to the same ethical responsibilities as the company; at the same time it means the company retains ownership of the research results, findings, and insights. This solution is, of course, extremely problematic for researchers who want to retain intellectual property of the algorithms or conclusions they develop from the data. It also often excludes the option of depositing the work into the public domain, which means it is unlikely to be reused for a public good. This solution benefits the company, not society, as the company often uses these tightly guarded analyses where no one else can.

While it is good the private companies have begun to investigate ways to share their data to address issues of the public good, it is clear that much work needs to be done to address the timely convenience of data, without having to be a company employee. The root of the problem is still the license agreements, which can take years to negotiate. Because the work to make these agreements takes time and money, and sharing data is not part of private companies essential business, they are often given the lowest priority. How then can we help private companies share data through ethical license agreements? Perhaps the best solution would be to have outside nonprofit organization help work with companies to set these agreements. This would help lessen the burden for private companies. Just as the World Bank negotiates license terms for all its researchers, this nonprofit could do that work for numerous special interest groups, from public health to criminal justice. The nonprofit organization established would be a great investment for funders, and ultimately private companies could also invest in the foundation as a form of philanthropy.

Data Philanthropy

There have been attempts to encourage private companies to see data sharing for a public good as a form of philanthropy and to make sharing data a more altruistic gesture. The idea of *data philanthropy* originated in 2011 with Robert Kirkpatrick of the United Nations Global Pulse, an initiative to accelerate the innovative use of big data for sustainable development and humanitarian action. Kirkpatrick suggests that corporations can find ways to share data for the public good, and in his article on the topic he mentions a few ways that can be done: he calls for a "data commons" to which companies can contribute data after is has been anonymized, and he reminds us of the idea of "strategic leaking" proposed by Michael Nelson of Microsoft's technology futures.[40] More generally, data philanthropy has come to involve donating data, whether by individuals or private companies.[41]

A great example of data philanthropy is the Data for Development (D4D) Challenge, put on by Orange, one of the largest telecoms in the world. Orange released its call detail record (CDR) in two African regions to see if the innovative minds of researchers around the world could think up creative ways to use this data as a public good. CDR data for the Ivory Coast and Senegal was released in 2013 and 2014 respectively, based on anonymized phone calls and SMS exchanges among millions of Orange's customers in those countries. According to articles detailing the project, "the datasets are: (a) antenna-to-antenna trace on an hourly basis, (b) individual trajectories for 50,000 customers for two-week time windows with antenna location information, (3) individual trajectories for 500,000 customers over the entire observation period with sub-prefecture location information, and (4) a sample of communication graphs for 5,000 customers."[42] Across the board, the challenge process dealt with issues essential to policy in Africa. The winners of the Senegal Orange Challenge in 2014–2015, for example, addressed topics of health, agricultural production, electrification planning, identified socio-demographics, electrification, poverty, and transportation.

CDR data has also been used to track malaria and Ebola, among other infectious diseases.[43] Although obtaining CDR data is really useful for natural disasters, companies often need to negotiate their terms in real time, and this is when ethical issues can arise: the telecoms often give out the data without considering the privacy of those *in* the data, as they believe their purpose is more important. McDonald mentioned this in his critique of the Ebola crisis. The Senegal Orange Challenge was held after the Ebola disaster had abated, which can also provide some insights, but having it during a crisis would certainly provide a great benefit as well. The timeliness of data acquisition is important, and this is why McDonald believes that pre-negotiated license terms are essential for using data during disasters.

Matt Stempeck, Director of Civic Technology at Microsoft, argues that the provision of data can make use of the International Charter on Space and Major Disasters (1999), which provides a clear protocol for satellite companies to give away data during a disaster. While Stempeck makes a great argument in favor of how private companies "could give back in meaningful ways" by providing "data for public good," especially when it comes to benefits for academic research and what he calls "responsive cities," he is less clear how this works in favor of companies, which are always interested in their bottom line, the good PR of philanthropy notwithstanding.[44] Traditional forms of philanthropy help companies save money, and while Stampeck provides some suggestions on how this could be done in data philanthropy, companies simply have not picked it up. One of the biggest reasons appears to be privacy concerns.

Mariarosaria Taddeo, Deputy Director of the Digital Ethics Lab at Oxford, writes that data philanthropy is more complicated than it seems on the face of it. There are ethical concerns regarding sharing private data even if it is for a public good. She argues that data can be used for the benefit of society, but she also cautions that it can be used in ways harmful to society; and she calls out the moral ambiguity that, while data philanthropy might have a moral *objective*, giving away data may not always be morally upright or ethical. "Data philanthropy can only maintain its promise to foster morally good outcomes—e.g., advancing scientific knowledge, improving policy-making processes and emergency-facing procedures—when combined with the right ethical principles."[45] Moreover, Taddeo writes, and I agree that, "more than natural resources, shared data are comparable to public goods. Like libraries, lighthouses or public parks, these data have the potential to foster the welfare of societies and of their members." Even with the call to use data for a public good during disaster events, the Data Pop Alliance (DPA) argues that CDR data is still not widely distributed even during crisis, and when it is, giving out CDRs in certain social political contexts might be good for the crisis but used against the society and individuals later.[46]

Since 2016, Bloomberg has held a conference called Data for Good Exchange. In addition, there has been no shortage of media coverage, conferences, and meetings on using data for a public good. Despite the hype around the topic, there are few examples to guide governments, NGOs, researchers, and the private companies themselves about how they might use big data to answer societal questions. Perhaps the biggest limitation is the clear incentive structure in place for companies who want to sell their data, and few guidelines in place for when companies want to overcome the risks involved in giving away their data. For companies to give away their data, making the process timely is rarely a high priority, which often means the data is no longer useful once it is acquired because it is out of date. And privacy is a barrier, too. The question thus remains: How do companies retain customer and citizen privacy as well as their core business model? This is an essential topic for research in the years to come, as data ownership by private companies will grow exponentially. We must invest time and money toward figuring out ways to use this data for public good ethically and responsibly, as we can't count on government regulation. Data Action asks that we identify new creative way to access and quantify data for the public good. Finding new ways to facilitate data sharing ethically and responsibly is part of that work, now and in the future.

CONCLUSION: IT'S HOW WE WORK WITH DATA THAT REALLY MATTERS

Data Action sets out to remind us that big data in its raw form cannot perform on its own; rather how data is transformed and operationalized can change the way we see the world. More specifically, data can be used for civic action and policy change by communicating with the data clearly and responsibly to expose the hidden patterns and ideologies to audiences inside and outside the policy arena. Communicating with data in this way requires the ability to ask the right questions, find or collect the appropriate data, analyze and interpret that data, and visualize the results in a way that can be understood by broad audiences. Combining these methods transforms data from a simple point on a map to a narrative that has meaning. Data is not often processed in this way because data analysts are often not familiar with the techniques that can be used to tell stories with the data ethically and responsibly.

Data Action seeks to provide guidance for using data toward the benefit of society, learning from the ways we have used data unethically in the past and illustrating ways we can use it more ethically and creatively in the future. Seven themes emerge for working with data for the public good, and these themes help to establish the principles for Data Action.

1. We must interrogate the reasons we want to use data and determine the potential for our work to do more harm than good.
2. Building teams to create narratives around data for action is essential for communicating the results effectively, but team collaboration also helps to make sure no harm is done to the people represented in the data itself.

3. Building data helps change the power dynamics inherent in controlling and using data, while also having numerous side benefits, such as teaching data literacy.

4. Coming up with unique ways to acquire, quantify, and model data can expose messages previously hidden from the public eye; however, we must expose ideas ethically, going back to the first principle above.

5. We must validate the work we do with data by literally observing the phenomenon on the ground and asking those it effects to interpret the results.

6. Sharing data is essential for communicating the need for policy change and generating a debate essential for that work. Data visualizations are effective at doing that.

7. We must remember that data are people, and we must do them no harm. Regulations help provide standards of practice for the use of data, but they often are not developed in line with technological change; therefore, we must seek to develop our own standards and call upon others to do the same. I present these principles in turn below.

The Purpose for Using Data Analytics Must Be Interrogated

Data has the same potential to harm society as it has to do good, and therefore it is essential that we interrogate the reasons for our own analysis and well as that of others. Data is an effective tool for policy action: it is good at convincing us of all kinds of things because it is equated with fact, and therefore rarely questioned. But as a tool, similar to a writer's pen or an artist's brush, data draws a picture of the world from the perspective of the person using it. Because we equate data with fact, we rarely argue against its picture; this also makes data particularly powerful and alarming for creating change in society. We must, therefore, be critical of our own analytics and that of others to make sure the work we do with data is not harmful to society.

The use of data analytics as evidence of change in society has a controversial history because it often reinforces the ideology of those in power. Harland Bartholomew's decades-long use of data was perhaps most telling example of this, because the blatant evidence he left behind proved he used his analytics to marginalize and segregate black communities in St. Louis and elsewhere. Data modeling techniques developed for cities after World War II often fell short because the analysts did not have sufficient expertise with urban issues to address the problems they were tackling. Nor could they judge whether the data at their disposal was incomplete or inaccurate. These early models show the hubris of data analysts practicing from the 1950s through the 1970s, who thought that computation alone could help the conditions of cities. This type of hubris still exists today with all the excitement about data and its ability to answer almost every situation, condition, or problem. But I argue that analysts must begin by asking policy questions of people with on-the-ground expertise, those who know the issue the best—and by believing ultimately that this collaboration will create smarter models.

As obvious as this may seem to some, data analytics often are not performed in this way. Let's look at what happened with Google Flu Trends, a model that uses Google search-term data to predict flu. Unfortunately, critiques of the model *by policy experts* were ignored, causing it to fail. Although Google fixed its mistakes, the public could no longer trust its results. Both then and now, excluding subject experts from the data analytics process leads to deceptive or inaccurate results; that ultimately leads to a mistrust of the outcomes and more failed models. Including policy experts in the process of innovating with data analytics is crucial.

Building Expert Teams Is Essential to Making Data Work for Policy Change

At the heart of all the work outlined in these pages is the idea that working collaboratively with policy experts, communities, and designers is essential to reduce the potential for analytics to guide us toward misleading, unethical, or inaccurate conclusions. But more importantly, building expert teams helps *communicate* the work. This is not to say we need to build consensus around the results of our data analytics, but rather encourage others—specialists, data analysts, and local or affected actors—to be involved in the project and critique the data analysis. Data analysis should include interactive feedback loops from people who understand the issue the best. Collaboration, then, allows us to better understand the policy questions we seek to answer, and experts can help guide the team through the nuances of policy issues and help improve data models. Different points of view also help to expose different perspectives and create a debate on the issues, which is essential for healthy government. Collaboration helps to build communities around an idea, and these communities help spread the results far and wide.

A good example of this is the work by the New York Civil Liberties Union (NYCLU) regarding New York City's stop-and-frisk policy. NYCLU created a larger community organization, Communities United for Police Reform (CPR), combining several organizations interested in race and policing. This new community set an agenda to employ stop-and-frisk data and thus to expose underlying prejudices in police behavior in New York City. Together the CPR organizers used their diverse networks to get the message out to the press, often in the form of data visualizations. The collaboration allowed the work to extend far and wide, so much so that it became a topic of the 2012 mayoral campaign.

The transparency developed through collaboration creates trusting relationships among the diverse actors as well. The work of the Digital Matatus project in Nairobi shows how sharing data created trust among matatu drivers and owners, policy experts, government officials, NGOs, and the public in Nairobi, and allowed the work to go beyond simply creating a map of semiformal transit: it changed how the system was planned. In the long run, collaboration saves time, as it helps to ensure the work is of high quality, will have an impact, can be trusted, and ultimately reaches the agenda of policy makers.

Building Data Changes Power Dynamics and Shapes Communities

Data Action makes a call for building data, which reminds us that even in the era of big data some data is missing or inaccessible—and sometimes purposely so. We must therefore fill the gaps in data so the voices of people, often on the margins of society, are not left unheard. Building data also creates a powerful voice for people whose perspective may be disregarded at the policy table. From the example of fracking in Pennsylvania to the measurement of air quality in Beijing, building data creates hard evidence that is difficult to argue against and can be used to change policy. Ultimately, building data shifts power dynamics as communities begin to control the narrative developed by data analytics and wield that power toward their needs and interests. The work gives them ammunition against big corporations and governments.

The narratives described in these pages show that building data has other benefits: it teaches data literacy, builds communities around shared ideas, and creates media buzz around topics by placing them on the policy agenda. Whether it be collecting data for OpenStreetMap or building air quality sensors; creating the measurement tools or collecting data in the field, interpreting and presenting the results build skills that are not only essential to the project but also valuable for our data-driven society: these projects teach participants to become data literate.

Data collection projects create new activist communities because diverse groups need to depend on one another to learn the tools needed to use data toward a shared goal. This was true for balloon mapping work performed during the 2010 *Deepwater Horizon* oil spill, when the data itself didn't matter as much as the process of engaging in it. The work brought diverse communities together to check the narrative of government released data about the extent of the oil spill. The media buzz around the innovative ways they created the data, balloons flying over restricted land, did a lot more to get the topic on the policy agenda then the results they developed. Using the media to share our topics is one way we use data to generate policy debate.

Quantify Ingeniously, but Remember Data Is Biased by Its Creator

While building data is a centuries-old strategy made easy by new technologies, innovations in technology also provide creative ways for us to access data that we never had before. Data can now be mined from websites or through private companies and databases using their openly available Application Programming Interfaces (APIs). These data sets can provide an essential resource in places where data is unavailable or tightly controlled. The work my team performed to measure the extent of ghost cities in China is a great example of this. Ghost cities, or neighborhoods with high rates of vacancy, symbolize risk in the Chinese real estate market; their location is tightly controlled by the Chinese central government, so much so that even local municipalities often do not have access to the information. The Ghost Cities project tapped into Chinese social media sites to gather data for the development of a model that identified these areas using the premise that healthy cities need access to amenities, and neighborhoods that lacked amenities were likely vacant. The artifact of the model not only helped to identify where these neighborhoods existed, but also helped to create potential solutions by pointing out the amenities that were deficient—and therefore investments could be made to improve the livability of the neighborhoods. Using data creatively can help expose the effects of policies: in this case Chinese strategies for economic development that would have otherwise been hard to conceptualize.

But when we use data in creative ways it is important to remember data holds the biases of its creators, because the data has been collected for a specific purpose. In the case of the Ghost Cities project the model didn't perform as well in older neighborhoods where amenities such as street vendors tended to be less formal and therefore were not registered on social media. This reminds us that data it never truly raw, it is collected—and the purpose for that collection can cause its use to produce misleading or inaccurate results. We must interrogate our data to expose these biases. This idea that data comes with its own values is not new for many data scientists, but is for many data novices who are grabbing their data from the web.

It's often impossible to remove data bias, but nonetheless important to acknowledge that biases exist—and sometimes even to exploit the biases to expose underlying value systems held in data collected by others. Investigating who or what is missing from data sets can be quite telling. My work with 311 calls in New York City provides an example: our team found that the places where people complained less about sanitation were also immigrant communities. That doesn't necessarily mean these communities were satisfied with the services; it is more likely related to a lack of knowledge about making 311 calls in the community, or to cultural norms around government responsibilities toward the public. As a result, we used the analysis as evidence to prove the need advertise the 311 hotline in languages other than English. Lack of data can also illustrate digital divides—areas that have fewer technologies can point to the need for the development of technology infrastructure.

Ground-Truthing Our Results with the People Data Analysis Affects

Using data in this way shows the importance of validating the results, or ground-truthing our data analytics, with the people represented in the data or by those who will be affected by its implications. This is why sharing data through visualizations is essential to Data Action: it allows people to see the implication of our analysis and critique the results without having to be data scientists. All the projects described in this book have sought the critique of those represented in the data, and through that work the outcomes of the models were transformed. Interactive visualizations, made possible by the web, allow sharing to be more transparent because people can explore the data on their own, and contribute their own insights to it. This process adds new intelligence to the project while also providing a way to ground-truth the results. Critical cartographers often argue that the one-dimensional perspective of maps limits our ability to judge the results presented, whereas interactive data displays help to overcome this disadvantage. This type of sharing was particularly useful in the Ghost Cities project, as the interactive visualization helped explain the model, allowed stakeholders to interact with the data, and ultimately helped generate a debate about ghost cities from a more informed perspective.

Data Brings Insights to the Public in Dynamic Ways

Beyond ground-truthing, sharing data is essential for Data Action because it brings the insights to broad publics, thereby helping to close knowledge gaps essential for making decisions for society. Therefore, when building our teams to work with data for policy change, we must include designers that can communicate our policy arguments through data narratives. A great example of this is the Million Dollar Blocks project, which used data developed by the criminal justice system to generate maps to expose, block by block, the extraordinary amounts of money we spend to incarcerate people in prisons outside of their communities, while at the same time in communities of poverty we spend little money to address the systemic reasons that contribute to imprisonment or recidivism, such as poor educational and economic opportunities or lack of reentry services. The red blocks that stand out against the dark background of the map help to highlight the work's alarming narrative. Using the city block as the tool to measure this phenomenon was as important a device as it was relatable. The Million Dollar Blocks maps didn't present a solution, but rather brought the issue to the public eye, allowing others to use the maps for their own advocacy.

This brings us full circle. Ever since William Playfair developed the first chart in 1786, visualizations have been effective tools for swaying the public because they retain the legitimacy of data. And yet that power can be used to elevate some and to marginalize others. So how, then, do we ensure the visualizations we make, or the algorithms and insights we generate from data are ethical? This is an important question, because no matter how we use data we are affecting people. I believe the narratives we create around data insights help guide how data visualizations will be used.

Let's take the example of the Brookes Slave Ship Map: an abolitionist created a narrative around the maps that illustrated the horrors of slavery, and they were received as such. Let's also use the example of the Home Owners' Loan Corporation (HOLC) redlining maps, now infamous for their marginalization of communities. It is often said that these maps were misappropriated to help disinvestment in communities of color. The maps marked areas of risk for investment, and their message was as clear as it was intentional—their bright red markings screamed, "Stay away!" The narratives generated around them were of risk and exclusion, not ones that set out to improve the conditions of those communities but rather to encourage investment in communities that were already doing well. Now, let's imagine if the narrative of these maps had been different: What if instead of excluding these communities from investment they sought to help improve their condition? What if the areas that needed investment to improve were marked in green instead of red? Could these same maps have helped to alleviate economic and racial divisions? How we use maps and discuss them changes the way they will be used by people outside the team that created them.

Data Are People, and We Must Do Them No Harm

It's not just the ethics of visualizations that are important to consider: within the exponential amounts of data we create everyday lies all of us—our thoughts and ideas posted on social media, all of our email interactions, our daily movements. Data are people, and we must do them no harm. For example, cell phone call detail record (CDR) data could unintentionally expose persecuted people during disaster events if that data were disseminated without regard to the proper precautions, or without determining who has access to the data and for what reason. These protections are often forgotten in the name of the public good. Some people want to remain hidden; this might be especially true in informal economies. In Nairobi we spoke to the matatu owners to ensure that mapping their stops and routes was something they supported, and would not expose them to harm. The matatus owners were in favor of generating the map in order to bring attention to their needs. But you can imagine how that would not be the case in other contexts where this type of infrastructure is criminalized even though people depend on it daily. It is essential that before we build data we make sure that the people represented in the data want their story told.

In the next decade, private institutions will have control over more data than governments, and in some countries that is already the case. While tons of data exist, the gap between who has access to data and who does not is widening. The massive amounts of data we now generate add up to a new, as yet unshaped and unregulated, infrastructure similar to electricity at the turn of the nineteenth century. All this new data amounts to a public good that should by definition be protected by governments. But most governments are underprepared to engage in this type of protection, and pointedly so. Congressional hearings about Facebook's involvement in the Cambridge Analytica scandal are a case in point. During the debate it was clear that many congressmen didn't understand that social media companies profit from the reuse of our data.

While the General Data Protection Regulations (GDPRs) have started to create more awareness about the mass amounts of data we now produce and consume, government regulations often lag behind technological advancements. It is therefore up to the public to self-regulate. This is a difficult task when regulations are not always in one's best interest, and data companies do not always perform the task of monitoring themselves well. And it's questionable whether we want governments to be responsible for protecting our data in the first place. Surveillance technologies are easy to abuse, and in dictatorships data can be a tool of oppression. Who, then, should protect our data? These are lingering questions that remain unsolved. We can nevertheless pledge to vigilantly interrogate the ethics of our data use and expose the misdeeds of those who use unethical data practices. Data ultimately contains the bias of those who control it; the answer, therefore, is to advocate for all types of political positions and use Data Action, as I have explained it in these pages, for the betterment of society.

ACKNOWLEDGMENTS

At the core of *Data Action*'s argument is the idea that creating change with data requires collaboration among diverse groups of people. Given this fact, it should come as no surprise that the number of people I need to thank amounts to "big data," but let me attempt to provide a representative sample.

First, I want to thank my extremely supportive colleagues at Massachusetts Institute of Technology (MIT). I am tremendously lucky to be working with such generous people. In particular, I want to thank Anne Sprin, who spent hours with me charting, chapter by chapter, the narrative I wanted to tell. She helped me find my editor and did some editing herself, and her faith in me meant more than I can express. Similarly, I want to thank Eran Ben-Joseph, who encouraged me to translate my ideas into a book that could have a real impact. Thanks also to Catherine D'Ignazio, who read an early version of the prospectus. I also want to thank the numerous researchers, students, and staff who worked with me in the Civic Data Design Lab at MIT. I had many editors along the way; Lisa Halliday and Megan Hustad assisted with the development of the prospectus and first chapters. A special thanks to Patricia (Patsy) Baudoin, the editor who helped me bring this book to the finish line. I am so grateful for Patsy's flexibility, questions, critiques, persistence, patience, and encouragement. Thanks to the numerous MIT students who helped with research, interviews, and the final production, including Dylan Halpren, Lily Bui, Karuna Mehta, and Chaewon Ahn. I also want to thank my editor at the MIT Press, Gita Devi Manaktala, and the external reviewers she chose to critique the first draft I submitted.

Each chapter of the book describes one of my projects in detail, so I want to thank my collaborators on those projects. I will start by thanking the Digital Matatus core team: Jacqueline (Jackie) M. Klopp, Peter Waiganjo Wagacha, Dan Orwa, and Adam White. It was through the work on Digital Matatus that I devised the Data Action method. It was Jackie who first invited me to work on a project in Nairobi in 2006, and after close to fifteen years of work, I consider Nairobi a second home, in large part due to that first invitation. Working with Jackie, a political scientist, has taught me the power of collaborating closely with people outside your field. Peter Waiganjo Wagacha and Dan Orwa, from the University of Nairobi's Department of Computer Science, taught me much about Kenya's people and technology. I would like to recognize the many students who have been involved in the Digital Matatus project at both the University of Nairobi and MIT: Wenfei Xu, Alexis Howland, Emily Eros, Elizabeth Resor, Rida Qadri, Kadeem Kahn, Stepehen Osoro, Collins Mundi, Derick Lung'aho, Jackson Mutua, Samuel Kariu, Peter Kamirir, and Carmelo Ignaccolo, among others. Thanks to Benjamin de la Pena, who funded the first grant for Digital Matatus while at the Rockefeller Foundation.

Ghost Cities is another project that helped me develop my Data Action method and I want to give a special word of thanks to Wenfei Xu, who worked on the project as a Senior Researcher in the Civic Data Design Lab. Thanks also to the Ghost Cities core research team, which included Zhekun Xiong, Changping Chen, Michael Foster, Shin-bin Tan, and Ege Ozgirin. Thanks also to Chaewon Ahn, who helped lead the Ghost Cities exhibition in Seoul, South Korea.

Thanks to the Associated Press for taking a risk and embarking on their first real-time data collection project during the Beijing Olympics. A special thanks goes to Siemond Chan, now at the *Wall Street Journal*. Without Siemond's unwavering support, this project would have never happened.

Beyond thanking my collaborators on the three main projects discussed in the book, I would like to thank such long-time collaborators as Christine Gaspar and Valeria Mogilevich from the Center for Urban Pedagogy, who worked on City Digits, and Elizabeth Currid-Halkett, with whom I worked on numerous projects, including the Geography of Buzz. Special thanks goes to Laura Kurgan, with whom I spent seven years co-directing the Spatial Information Design Lab (SIDL) at Columbia University's Graduate School of Architecture, Planning, and Preservation (GSAPP). We also collaborated on the Architecture and Justice project discussed in this book. Laura is a mentor, colleague, and friend, and I am grateful for her advice and critiques. I also want to thank Mark Wigley, the Dean of GSAPP during my time at the Spatial Information Design Lab (SIDL). His support gave me the platform I needed to test my ideas and help formulate my practice. I also want to thank Georgia Bullen, Juan Francisco Saldarriaga, Erica Deahl, and Hayretin Gunc, senior researchers in both labs who worked on numerous projects covered in *Data Action.*

Finally, I need to thank all my wonderful supportive friends, colleagues, and family. This book is dedicated to all of you who have been so patient with me and encouraged me along the way. It is because of your support that you hold this book in your hands—I could not have written it without you. Special thanks to my father, whose work as journalist inspired me to pursue civic service, and to my stepmother, Marcia Myers, also a journalist, who took a keen editorial eye to many of the pages.

Given that this book explores in detail the projects that have helped define my Data Action methodology, some of the work described herein has been written about elsewhere. I cite those publications throughout; however, I would like to specifically mention the longer format case studies found in chapters 2, 3, and 4. Findings from the Beijing Olympics project, discussed in chapter 2, were included in "Beijing Air Tracks: Tracking Data for Good," in *Accountability Technologies Tools for Asking Hard Questions*, edited by Dietmar Offenhuber and Katja Schechtner (Vienna: AMBRA, 2013). The Ghost Cities project, discussed in chapter 3, was previously written about in Sarah Williams, Wenfei Xu, Mike Foster, Chanping Chen, and Shin Bin Tan, "Ghost Cities of China: Identifying Urban Vacancy through Social Media Data," *Cities* 94 (2019): 275–285. The Digital Matatus project, covered in chapter 4, was written about in Sarah Williams, Adam White, Peter Waiganjo, Daniel Orwa, and Jacqueline Klopp, "The Digital Matatu Project: Using Cell Phones to Create an Open Source Data for Nairobi's Semi-formal Bus System," *Journal of Transport Geography* 49 (2015): 39–51, among several other journals. Although the discussion of these projects in *Data Action* has been modified to illustrate how they turned data into action, it is worth noting how the work has been described elsewhere.

NOTES

Introduction

1. Sherry Arnstein, "A Ladder of Citizen Participation," *Journal of the American Institute of Planners* 35, no. 4 (1969): 216–224.

2. Lisa Gitelman, *"Raw Data" Is an Oxymoron* (Cambridge, MA: MIT Press, 2013).

3. Cathy O'Neil, *Weapons of Math Destruction: How Big Data Increases Inequality and Threatens Democracy* (New York: Broadway Books, 2017), p. 3.

4. Safiya Umoja Noble, *Algorithms of Oppression: How Search Engines Reinforce Racism* (New York: New York University Press, 2018).

5. James C. Scott, *Seeing like a State: How Certain Schemes to Improve the Human Condition Have Failed* (New Haven: Yale University Press, 1998).

6. David Reinsel, John Gantz, and John Rydning, "The Digitization of the World from Edge to Core," accessed January 25, 2019, https://www.seagate.com/files/www-content/our-story/trends/files/idc-seagate-dataage-whitepaper.pdf.

7. Carl Pierre, "Why You Need to Know What a Zettabyte Is [Infographic]," accessed January 3, 2020, https://www.americaninno.com/dc/why-you-need-to-know-what-a-zetabyte-is-infographic/.

8. Nick Couldry, *The Costs of Connection: How Data Is Colonizing Human Life and Appropriating It for Capitalism* (Stanford, CA: Stanford University Press, 2019).

9. Carpenter v. United States, 585 U.S. ___ (2018), https://www.supremecourt.gov/opinions/17pdf/16-402diff_1oc3.pdf.

Chapter 1

1. James C. Scott, *Seeing like a State: How Certain Schemes to Improve the Human Condition Have Failed* (New Haven: Yale University Press, 1998), 28.

2. Martin Bulmer et al., *The Social Survey in Historical Perspective, 1880–1940* (Cambridge: Cambridge University Press, 1991).

3. Etienne Balazs, *Chinese Civilization and Bureaucracy: Variations on a Theme* (New Haven: Yale University Press, 1967); Susan E. Alcock et al., *Empires: Perspectives from Archaeology and History* (Cambridge, UK: Cambridge University Press, 2001).

4. John S. Bowman, *Columbia Chronologies of Asian History and Culture* (New York: Columbia University Press, 2000).

5. Playfair, William, *Commercial and Political Atlas and Statistical Breviary (Original Version Was Published in 1786)* (Cambridge: Cambridge University Press, 2005), 2.

6. Paul Beynon-Davies, "Significant Threads: The Nature of Data," *International Journal of Information Management* 29, no. 3 (June 1, 2009): 170–188, https://doi.org/10.1016/j.ijinfomgt.2008.12.003.

7. Martin Bulmer, Kevin Bales, and Kathryn Kish Sklar, *The Social Survey in Historical Perspective, 1880–1940* (Cambridge, UK: Cambridge University Press, 1991).

8. "Off the Map," *Economist*, November 13, 2014, https://www.economist.com/international/2014/11/13/off-the-map.

9. UN IEAG, "A World That Counts–Mobilising the Data Revolution for Sustainable Development," *New York: United Nations*, 2014.

10. "SDGs: Sustainable Development Knowledge Platform," accessed July 14, 2019, https://sustainabledevelopment.un.org/sdgs.

11. Margo J. Anderson, *The American Census: A Social History, Second Edition*, 2nd edition (New Haven: Yale University Press, 2015).

12. Ibid.

13. Emily Baumgaertner, "Despite Concerns, Census Will Ask Respondents if They Are U.S. Citizens," *New York Times*, March 28, 2018, https://www.nytimes.com/2018/03/26/us/politics/census-citizenship-question-trump.html.

14. Emily Badger, "A Census Question That Could Change How Power Is Divided in America," *New York Times*, August 2, 2018, https://www.nytimes.com/2018/07/31/upshot/Census-question-citizenship-power.html.

15. Michael Wines, "Deceased G.O.P. Strategist's Hard Drives Reveal New Details on the Census Citizenship Question," *New York Times*, May 30, 2019, https://www.nytimes.com/2019/05/30/us/census-citizenship-question-hofeller.html.

16. Wines, "Deceased G.O.P. Strategist's Hard Drive"; Richard L. Hasen, "New Memo Reveals Census Question Was Added to Boost White Voting Power," *Slate*, accessed June 26, 2019, https://slate.com/news-and-politics/2019/05/census-memo-supreme-court-conservatives-white-voters-alito.html.

17. Hasen, "New Memo Reveals Census Question"

18. David Daley, "The Secret Files of the Master of Modern Republican Gerrymandering," *New Yorker*, September 6, 2019, https://www.newyorker.com/news/news-desk/the-secret-files-of-the-master-of-modern-republican-gerrymandering.

19. Ibid.

20. Justin Levitt, "A Citizen's Guide to Redistricting," SSRN Scholarly Paper (Rochester, NY: Social Science Research Network, June 28, 2008), https://papers.ssrn.com/abstract=1647221.

21. Ibid.

22. Ibid.

23. Map Data: Texas Department of Transportation Open Data, TxDOT Open Data Portal (Texas,2019),https://gis-txdot.opendata.arcgis.com/datasets/texas-us-house-districts?geometry =-334.512%2C-52.268%2C334.512%2C52.268; Department of Transportation Open Data, OpenStreetMap contributors, "Planet Dump," https://planet.osm.org," 2017; "Federal Court Rules Three Texas Congressional Districts Illegally Drawn," *NPR*, accessed January 30, 2019, https://www.npr.org/sections/thetwo-way/2017/03/11/519839892/federal-court-rules -three-texas-congressional-districts-illegally-drawn.

24. Patrick Joyce, *The Rule of Freedom: Liberalism and the Modern City* (London: Verso, 2003).

25. London and the Industrial Revolution," *Londontopia* (blog), accessed June 27, 2019, https://londontopia.net/history/london-industrial-revolution/.

26. Bulmer et al., *The Social Survey in Historical Perspective, 1880–1940*.

27. It should be noted that these early statistical societies were largely driven by people trained as sociologists; therefore, they were much more interested in using statistics to understand social issues.

28. Bulmer et al., *The Social Survey in Historical Perspective, 1880–1940*.

29. About Parliament, "1842 Sanitation Report," https://www.parliament.uk/about/living -heritage/transformingsociety/livinglearning/coll-9-health1/health-02/1842-sanitary -report-leeds/.

30. Bulmer et al., *The Social Survey in Historical Perspective, 1880–1940*.

31. Joyce, *The Rule of Freedom*; Jeffrey G. Williamson, *Coping with City Growth during the British Industrial Revolution* (Cambridge: Cambridge University Press, 2002).

32. Anne Hardy, "'Death Is the Cure of All Diseases': Using the General Register Office Cause of Death Statistics for 1837–1920," *Social History of Medicine* 7, no. 3 (December 1, 1994): 472–492, https://doi.org/10.1093/shm/7.3.472.

33. Joyce, *The Rule of Freedom*.

34. IDH Shepherd, "Mapping the Poor in Late-Victorian London: A Multi-Scale Approach," in *Getting the Measure of Poverty* (London: Aldershot, Ashgate, 1999), 148–176.

35. Charles Booth, Richard M. Elman, and Albert Fried, *Charles Booth's London* (New York: Pantheon Books, 1968).

36. Booth, Elman, anad Fried, *Charles Booth's London*.

37. Ibid.

38. Joyce, *The Rule of Freedom*.

39. Thomas Osborne and Nikolas Rose, "Spatial Phenomenotechnics: Making Space with Charles Booth and Patrick Geddes," *Environment and Planning D: Society and Space* 22, no. 2 (April 1, 2004): 209–228, https://doi.org/10.1068/d325t.

40. Vaughan, Laura, "The Charles Booth Maps as a Mirror to Society," *Mapping Urban Form and Society* (blog), October 3, 2017, https://urbanformation.wordpress.com/2017/10/03/the-charles-booth-maps-as-a-mirror-to-society/.

41. R. Vladimir Steffel, "The Boundary Street Estate: An Example of Urban Redevelopment by the London County Council, 1889–1914," *Town Planning Review* 47, no. 2 (1976): 161–173.

42. Ben Gidley, *The Proletarian Other: Charles Booth and the Politics of Representation* (London: Goldsmiths University of London, 2000).

43. Amy E. Hillier, "Redlining and the Home Owners' Loan Corporation," *Journal of Urban History* 29, no. 4 (May 1, 2003): 394–420, https://doi.org/10.1177/0096144203029004002.

44. Greg Miller, "1885 Map Reveals Vice in San Francisco's Chinatown and Racism at City Hall," *Wired*, September 30, 2013, https://www.wired.com/2013/09/1885-map-san-francisco-chinatow/.

45. Shah, *Contagious Divides*.

46. Mary Jo Deegan, "W. E. B. Du Bois and the Women of Hull-House, 1895–1899," *American Sociologist* 19, no. 4 (December 1, 1988): 301–311, https://doi.org/10.1007/BF02691827.

47. Deegan, "W. E. B. Du Bois and the Women of Hull-House."

48. Ibid.

49. Elliott M. Rudwick, "WEB Du Bois and the Atlanta University Studies on the Negro," in *WEB Du Bois* (Routledge, 2017), 63–74.

50. Bulmer, Martin, *The Chicago School of Sociology: Institutionalization, Diversity, and the Rise of Sociological Research* (Chicago: University of Chicago Press, 1986); Robert E. Park and Ernest W. Burgess, *The City* (Chicago: University of Chicago Press, 2012).

51. Davarian L. Baldwin, "Chicago's 'Concentric Circles': Thinking through the Material History of an Iconic Map," in *Many Voices, One Nation: Material Culture Reflections on Race and Migration in the United States*, ed. Margaret Salazar-Porzio, Joan Fragaszy Troyano, and Lauren Safranek (Washington, DC: Smithsonian Institution, 2017), 179–191.

52. David C. Hammack and Stanton Wheeler, *Social Science in the Making: Essays on the Russell Sage Foundation, 1907–1972* (Russell Sage Foundation, 1995).

53. Jon A. Peterson, *The Birth of City Planning in the United States, 1840–1917* (Baltimore: Johns Hopkins University Press, 2003). "Having outgrown the City Beautiful movement, its leaders sought to deepen their expertise, enhance their legal powers, refine their methodology, and above all make planning work. Substantively, they laid claims to the control of private property, most notably through zoning. Procedurally, they adopted the social survey as a prerequisite to the plan making, while institutionally, they made the city planning commission their preferred mechanism for attaining results" (p. 263).

54. Andreas Faludi, *A Reader in Planning Theory* (New York: Pergamon Press, 1973); A. J. Harrison, *Economics and Land Use Planning* (London: Croom Helm, 1977); Mark Pennington, "A Hayekian Liberal Critique of Collaborative Planning," *Planning Futures: New Directions for Planning Theory*, 2002, 187–208.

55. Eldridge Lovelace, *Harland Bartholomew: His Contribution to American Urban Planning* (Urbana: Department of Urban and Regional Planning, University of Illinois, 1993).

56. Richard Rothstein, "The Making of Ferguson," *Journal of Affordable Housing and Community Development Law* 24 (2016 2015): 165–204.

57. Harland Bartholomew, *The Zone Plan*, (St. Louis, MO: Nixon-Jones Printing Co., 1919), p. 82, https://catalog.hathitrust.org/Record/000342898.

58. Rothstein, "The Making of Ferguson," 174.

59. Rothstein, 174.

60. Rothstein, 175.

61. Lovelace, *Harland Bartholomew*.

62. Harland Bartholomew, *Urban Land Uses* (Cambridge, MA: Harvard University Press, 1932).

63. Hillier, "Redlining and the Home Owners' Loan Corporation."

64. Ibid.

65. Raymond A. Mohl, "Stop the Road: Freeway Revolts in American Cities," *Journal of Urban History* 30, no. 5 (July 2004): 674–706, https://doi.org/10.1177/0096144204265180. See also: Miami City Planning and Zoning Board, "The Miami Long Range Plan: Report on Tentative Plan for Trafficways" (Miami, 1955); Wilbur Smith and Associates, "A Major Highway Plan for Metropolitan Dade County, Florida" (New Haven, CT: Prepared for State Road Department of Florida and Dade County Commission, 1956); Smith and Associates, "Alternates for Expressways: Downtown Miami, Dade County, Florida" (New Haven, CT, 1962).

66. Mohl, "Stop the Road." See also: Miami City Planning and Zoning Board, "The Miami Long Range Plan"; Wilbur Smith and Associates, "A Major Highway Plan for Metropolitan Dade County, Florida"; Smith and Associates, "Alternates for Expressways."

67. Mohl, "Stop the Road."

68. Peter Hall, "The Turbulent Eighth Decade: Challenges to American City Planning," *Journal of the American Planning Association* 55, no. 3 (September 30, 1989): 275–282, https://doi.org/10.1080/01944368908975415.

69. Roberta Brandes Gratz, *The Battle for Gotham: New York in the Shadow of Robert Moses and Jane Jacobs* (New York: Nation Books, 2010).

70. Gratz, *The Battle for Gotham*; Jane Jacobs, *The Death and Life of Great American Cities* (New York: Random House, 1961).

71. Sherry Arnstein, "A Ladder of Citizen Participation," *Journal of the American Institute of Planners* 35. no. 4 (1969): 216–224.

72. John Forester, *Planning in the Face of Power* (Oakland: University of California Press, 1988).

73. James J. Glass, "Citizen Participation in Planning: The Relationship Between Objectives and Techniques," *Journal of the American Planning Association* 45, no. 2 (April 1979): 180–189, https://doi.org/10.1080/01944367908976956.

74. John Forester, *The Deliberative Practitioner: Encouraging Participatory Planning Processes* (Cambridge, MA: MIT Press, 1999).

75. Kevin Lynch, *The Image of the City* (Cambridge, MA: MIT Press, 1960).

76. Jennifer S. Light, *From Warfare to Welfare: Defense Intellectuals and Urban Problems in Cold War America* (Baltimore: Johns Hopkins University Press, 2003).

77. Trevor J. Barnes, "Geography's Underworld: The Military–Industrial Complex, Mathematical Modelling and the Quantitative Revolution," *Geoforum* 39, no. 1 (January 2008): 3–16, https://doi.org/10.1016/j.geoforum.2007.09.006.

78. Barnes, "Geography's Underworld."

79. Peter Hall, "The Turbulent Eighth Decade: Challenges to American City Planning," *Journal of the American Planning Association* 55, no. 3 (1989): 275–282, https://doi.org/10.1080/01944368908975415.

80. Forester, *Planning in the Face of Power*.

81. Light, *From Warfare to Welfare*.

82. Mark Vallianatos, "Uncovering the Early History of the 'Big Data' and the 'Smart City' in LA," *Boom*, https://boomcalifornia.com/2015/06/16/uncovering-the-early-history-of-big-data-and-the-smart-city-in-la/.

83. Richard L Meier, *A Communications Theory of Urban Growth* (Cambridge, MA: MIT Press, 1962).

84. Anthony M. Townsend, *Smart Cities: Big Data, Civic Hackers, and the Quest for a New Utopia* (New York: Norton, 2013).

85. Ibid.

86. James N. Gray, David Pessel, and Pravin P. Varaiya, "A Critique of Forrester's Model of an Urban Area," *IEEE Transactions on Systems, Man, and Cybernetics*, no. 2 (1972): 139–144.

87. Michael Batty and Paul M. Torrens, "Modelling and Prediction in a Complex World," special issue, *Complexity and the Limits of Knowledge* in *Futures*, 37, no. 7 (September 1, 2005): 745–766, https://doi.org/10.1016/j.futures.2004.11.003.

88. Douglass B. Lee, "Requiem for Large-Scale Models," *Journal of the American Institute of Planners* 39, no. 3 (May 1973): 163–178, https://doi.org/10.1080/01944367308977851.

89. Light, *From Warfare to Welfare*.

90. Rob Kitchin, "The Real-Time City? Big Data and Smart Urbanism," *GeoJournal* 79, no. 1 (February 1, 2014): 1–14, https://doi.org/10.1007/s10708-013-9516-8.

91. Townsend, *Smart Cities*.

92. John Pickles, "Arguments, Debates, and Dialogues: The GIS–Social Theory Debate and the Concern for Alternatives," in Paul. A. Longley, Michael F. Goodchild, D. J. MacGuire, and David W. Rhind (eds.), *Geographical Information Systems: Principles, Techniques, Applications, and Management* 2nd ed. (New York: John Wiley and Sons, 2005), 12; Francis Koti and Daniel Weiner, "(Re) Defining Peri-Urban Residential Space Using Participatory GIS in Kenya," *Electronic Journal of Information Systems in Developing Countries* 25, no. 1 (June 2006): 1–12, https://doi.org/10.1002/j.1681-4835.2006.tb00169.x.

93. John Pickles, *Ground Truth: The Social Implications of Geographic Information Systems* (Guilford Press, 1995); Trevor Harris and Daniel Weiner, "Empowerment, Marginalization, and 'Community-Integrated' GIS," *Cartography and Geographic Information Systems* 25, no. 2 (1998): 67–76, https://doi.org/10.1559/152304098782594580; Sarah A Elwood, "GIS Use in Community Planning: A Multidimensional Analysis of Empowerment," *Environment and Planning A* 34, no. 5 (May 2002): 905–922, https://doi.org/10.1068/a34117.

94. Eric Gordon, Steven Schirra, and Justin Hollander, "Immersive Planning: A Conceptual Model for Designing Public Participation with New Technologies," *Environment and Planning B: Planning and Design* 38, no. 3 (June 2011): 505–519, https://doi.org/10.1068/b37013.

95. Sarah Elwood and Helga Leitner, "GIS and Spatial Knowledge Production for Neighborhood Revitalization: Negotiating State Priorities and Neighborhood Visions," *Journal of Urban Affairs* 25, no. 2 (2003): 139–157, https://doi.org/10.1111/1467-9906.t01-1-00003; Keiron Bailey and Ted Grossardt, "Toward Structured Public Involvement: Justice, Geography and Collaborative Geospatial/Geovisual Decision Support Systems," *Annals of the Association of American Geographers* 100, no. 1 (2010): 57–86, https://doi.org/10.1080/00045600903364259.

96. Janis B. Alcorn et al, "Keys to Unleash Mapping's Good Magic," *PLA Notes* 39, no. 2 (2000): 10–13; Giacomo Rambaldi, "Who Owns the Map Legend?," *URISA Journal* 17, no. 1 (2005): 5–13.

97. David S. Sawicki and William J. Craig, "The Democratization of Data: Bridging the Gap for Community Groups," *Journal of the American Planning Association* 62, no. 4 (December 31, 1996): 512–523, https://doi.org/10.1080/01944369608975715; Rina Ghose, "Politics of Scale and Networks of Association in Public Participation GIS," *Environment and Planning A* 39, no. 8 (2007): 1961–1980; Wendy A Kellogg, "From the Field: Observations on Using GIS to Develop a Neighborhood Environmental Information System for Community-Based Organizations" 11, no. 1 (1999): 19; Elwood, "GIS Use in Community Planning."

98. "IBM Smarter Cities—Future Cities—United States," August 3, 2016, http://www.ibm.com/smarterplanet/us/en/smarter_cities/overview/.

99. Natasha Singer, "IBM Takes 'Smarter Cities' Concept to Rio de Janeiro," *New York Times*, March 3, 2012, https://www.nytimes.com/2012/03/04/business/ibm-takes-smarter-cities-concept-to-rio-de-janeiro.html.

100. Pete Swabey, "IBM, Cisco and the Business of Smart Cities," *Information Age*, February 23, 2012, https://www.information-age.com/ibm-cisco-and-the-business-of-smart-cities-2087993/.

101. "The Data Deluge," *Economist*, February 25, 2010, https://www.economist.com/leaders/2010/02/25/the-data-deluge.

102. "Microsoft CityNext: Coming to a City Near You," Stories, July 10, 2013, https://news.microsoft.com/2013/07/10/microsoft-citynext-coming-to-a-city-near-you/.

103. Stephen Goldsmith and Susan Crawford, *The Responsive City: Engaging Communities Through Data-Smart Governance* (Hoboken, NJ: John Wiley & Sons, 2014).

104. Susan Fainstein and James DeFilippis, *Readings in Planning Theory* (Hoboken, NJ: John Wiley & Sons, 2015).

Chapter 2

1. The term "citizen" here should be understood broadly: "citizen data" might in fact be collected by noncitizens, resident aliens, and undocumented immigrants in addition to citizens proper, any of whom may become involved in an effort to deploy their data findings to improve the lot of their communities.

2. Reinsel, Gantz, and Rydning, "The Digitization of the World From Edge to Core," accessed January 25, 2019, https://www.seagate.com/files/www-content/our-story/trends/files/idc-seagate-dataage-whitepaper.pdf.

3. Robert Chambers, "Participatory Mapping and Geographic Information Systems: Whose Map? Who Is Empowered and Who Disempowered? Who Gains and Who Loses?," *Electronic Journal of Information Systems in Developing Countries* 25, no. 1 (2006): 1–11; Wen Lin, "Counter-cartographies," *Introducing Human Geographies*, 2013, 215.

4. Paulo Freire, *Pedagogy of the Oppressed* (New York: Bloomsbury Publishing USA, 2018).

5. Lin, "Counter-cartographies."

6. Robert Chambers, "The Origins and Practice of Participatory Rural Appraisal," *World Development* 22, no. 7 (July 1, 1994): 953–969, https://doi.org/10.1016/0305-750X(94)90141-4.

7. Raymond A. Mohl, "Stop the Road: Freeway Revolts in American Cities," *Journal of Urban History* 30, no. 5 (2004): 674–706.

8. For more information on the various terms see: J. Brian Harley, "Maps, Knowledge, and Power," *Geographic Thought: A Praxis Perspective*, 2009, 129–148; Jeremy W. Crampton and John Krygier, "An Introduction to Critical Cartography," *ACME: An International E-Journal for Critical Geographies* 4, no. 1 (2006): 11–33; Sarah Elwood, "Critical Issues in Participatory GIS: Deconstructions, Reconstructions, and New Research Directions," *Transactions in GIS* 10, no. 5 (2006): 693–708; and Daniel Sui, Sarah Elwood, and Michael Goodchild, *Crowdsourcing Geographic Knowledge: Volunteered Geographic Information (VGI) in Theory and Practice* (New York: Springer Science & Business Media, 2012).

9. Jeffrey A. Burke, Deborah Estrin, Mark Hansen et al., "Participatory Sensing," UCLA Center for Embedded Network Sensing, 2006, https://escholarship.org/uc/item/19h777qd.

10. Deborah Estrin et al., "Participatory Sensing: Applications and Architecture [Internet Predictions]," *IEEE Internet Computing* 14, no. 1 (2010): 12–42.

11. Eric Paulos, Richard J. Honicky, and Elizabeth Goodman, "Sensing Atmosphere," *Human-Computer Interaction Institute*, 2007, 203.

12. Pengfei Zhou, Yuanqing Zheng, and Mo Li, "How Long to Wait?: Predicting Bus Arrival Time with Mobile Phone Based Participatory Sensing," in *Proceedings of the 10th International Conference on Mobile Systems, Applications, and Services* (ACM, 2012), 379–392.

13. Sasank Reddy, Deborah Estrin, and Mani Srivastava, "Recruitment Framework for Participatory Sensing Data Collections," in *International Conference on Pervasive Computing* (New York: Springer, 2010), 138–155.

14. Michael F. Goodchild, "Citizens as Sensors: The World of Volunteered Geography," *GeoJournal* 69, no. 4 (2007): 211–221.

15. Pascal Neis and Alexander Zipf, "Analyzing the Contributor Activity of a Volunteered Geographic Information Project—The Case of OpenStreetMap," *ISPRS International Journal of Geo-Information* 1, no. 2 (September 2012): 146–165, https://doi.org/10.3390/ijgi1020146.

16. The first Free and Open Source Software for Geoinformatics (FOSS4G) conference was also held in 2006 and has become very popular as well.

17. Open-source software by definition is software that one can freely access and contribute source code. The community that grew around OpenStreetMap and FOSS4G made being open-source cool in the world of computer cartography. For instance, when I attended the 2018 FOSS4G conference in Boston I noticed many commercial mapping companies, including ArcGIS, showed up because they wanted to be branded as open-source.

18. Maureen Fan, "Gray Wall Dims Hopes of 'Green' Games; China Has Vowed to Curb Pollution Before '08 Olympics, but Its Secrecy Is Feeding Skepticism," *Washington Post*, October 16, 2007.

19. Jonathan Watts, "China Prays for Olympic Wind as Car Bans Fail to Shift Beijing Smog," *Guardian*, August 21, 2007.

20. Jim Yardley, "Beijing's Olympic Quest: Turn Smoggy Sky Blue," *New York Times*, December 29, 2007.

21. Maureen Fan, "If That Doesn't Clear the Air . . . China, Struggling to Control Smog, Announces 'Just-in-Case' Plan," *Washington Post*, July 31, 2008.

22. Patrick L. Kinney et al., "Airborne Concentrations of PM (2.5) and Diesel Exhaust Particles on Harlem Sidewalks: A Community-Based Pilot Study," *Environmental Health Perspectives* 108, no. 3 (2000): 213.

23. Incite! and Women of Color Against Violence, *The Revolution Will Not Be Funded: Beyond the Non-Profit Industrial Complex* (South End Press, 2007).

24. Roy M. Harrison and Jianxin Yin, "Particulate Matter in the Atmosphere: Which Particle Properties Are Important for Its Effects on Health?," *Science of the Total Environment* 249, no. 1 (April 17, 2000): 85–101, https://doi.org/10.1016/S0048-9697(99)00513-6.

25. Jiming Hao and Litao Wang, "Improving Urban Air Quality in China: Beijing Case Study," *Journal of the Air & Waste Management Association* 55, no. 9 (2005): 1298–1305.

26. The carbon monoxide sensor measured CO levels using chemical reactions. Chemical-based sensors exhibit degraded levels of accuracy the longer they are in use because the chemical agent in the sensor degrades. We purchased new carbon monoxide sensors from LASCAR and calibrated these sensors before leaving for Beijing.

27. There is some discussion in the scientific community about the accuracy of optical sensors versus mass-based sensors for measuring particulate matter. Particulate matter sensors that use optical technology measure the shadows of fine particles as air passes through a light component. Mass-based devices pump air into a filter bag, and once the particles are trapped, the bags are sent to a lab to be measured and analyzed for composition and size. After weighing the arguments for both, we purchased an optical sensor developed by MICRODUST, at the time considered one of the best optical sensors in the field. While there are benefits to using filter bags, including the analysis of particle composition, they would not have allowed us

to provide real-time reporting. We decided that when calibrated, real-time optical sensors would deliver the best estimate of the amount of particulate matter in Beijing's air. See Pranas Baltrenas and Mindaugas Kvasauskas. "Experimental Investigation of Particle Concentration Using Mass and Optical Methods." *Journal of Environmental Engineering and Landscape Management* 13, no. 2 (2005): 57–64.

28. The sensors were calibrated for air quality conditions in Beijing.

29. Although the marathons were officially run on different dates, our experiments were run on the same day—August 24, 2008.

30. Fan, "If That Doesn't Clear the Air."

31. Natasha Vicens, "How One Resident near Fracking Got the EPA to Pay Attention to Her Air Quality," *PublicSource*, December 15, 2016, https://www.publicsource.org/how-one-resident-near-fracking-got-the-epa-to-pay-attention-to-her-air-quality/.

32. Jennifer Gabrys, *Program Earth: Environmental Sensing Technology and the Making of a Computational Planet*, vol. 49 (Minneapolis: University of Minnesota Press, 2016).

33. Jeremy W. Peters, "BP and Officials Block Some Coverage of Gulf Oil Spill," *New York Times*, June 9, 2010, https://www.nytimes.com/2010/06/10/us/10access.html.

34. Patrik Jonsson, "Gulf Oil Spill: Al Gore Slams BP for Lack of Media Access," *Christian Science Monitor*, June 15, 2010, https://www.csmonitor.com/USA/Politics/2010/0615/Gulf-oil-spill-Al-Gore-slams-BP-for-lack-of-media-access.

35. Dorothy L. Hodgson and Richard A. Schroeder, "Dilemmas of Counter-Mapping Community Resources in Tanzania," *Development and Change* 33, no. 1 (2002): 79–100, https://doi.org/10.1111/1467-7660.00241.

36. The Crowd & The Cloud, "DIY Science for the People: Q&A with Public Lab's Jeff Warren," *Medium* (blog), March 27, 2017, https://medium.com/@crowdandcloud/diy-science-for-the-people-q-a-with-public-labs-jeff-warren-c45b6689a228.

37. Anya Groner, "Healing the Gulf with Buckets and Balloons," *Guernica*, September 8, 2016, https://www.guernicamag.com/anya-groner-healing-the-gulf-with-buckets-and-balloons/.

38 Public Lab Contributors, "Balloon Mapping Kit," n.d., https://store.publiclab.org/collections/featured-kits/products/mini-balloon-mapping-kit?variant=48075678607.

39. Ibid.

40. Ibid.

41. Paolo Cravero, "Mapping for Food Safety," International Institute for Environment and Development, December 21, 2015, https://www.iied.org/mapping-for-food-safety.

42. Azby Brown et al., "Safecast: Successful Citizen-Science for Radiation Measurement and Communication after Fukushima," *Journal of Radiological Protection* 36, no. 2 (2016): S82.

43. "First Safecast Mobile Recon," *Safecast* (blog), April 24, 2011, https://blog.safecast.org/2011/04/first-safecast/.

44. "Hands-On Projects for Curious Minds," n.d., KitHub homepage, https://kithub.cc/.

45. Jeff Howe, "The Rise of Crowdsourcing," *Wired*, June 1, 2006, https://www.wired.com/2006/06/crowds/.

46. Ibid.

47. Alek Felstiner, "Working the Crowd: Employment and Labor Law in the Crowdsourcing Industry," *Berkeley Journal of Employment and Labor Law* 32 (2011): 143–204; Alana Semuels, "The Internet Is Enabling a New Kind of Poorly Paid Hell," *Atlantic*, January 23, 2018, https://www.theatlantic.com/business/archive/2018/01/amazon-mechanical-turk/551192/.

48. Brown et al., "Safecast."

49. Jonathan Watts, "Almost Four Environmental Defenders a Week Killed in 2017," *Guardian*, February 2, 2018, https://www.theguardian.com/environment/2018/feb/02/almost-four-environmental-defenders-a-week-killed-in-2017.

50. Dietmar Offenhuber and Carlo Ratti, *Waste Is Information: Infrastructure Legibility and Governance* (Cambridge, MA: MIT Press, 2017).

51. Sarah Crean, "While Quietly Improving, City's Air Quality Crisis Quietly Persists," *Gotham Gazette*, June 19, 2014, http://www.gothamgazette.com/government/5111-while-improving-quiet-crisis-air-quality-persists-new-york-city-asthma-air-pollution.

52. Zev Ross et al., "Spatial and Temporal Estimation of Air Pollutants in New York City: Exposure Assignment for Use in a Birth Outcomes Study," *Environmental Health* 12 (June 27, 2013): 51, https://doi.org/10.1186/1476-069X-12-51.

53. Jane E. Clougherty and Laura D. Kubzansky, "A Framework for Examining Social Stress and Susceptibility to Air Pollution in Respiratory Health," *Environmental Health Perspectives* 117, no. 9 (September 1, 2009): 1351–1358, https://doi.org/10.1289/ehp.0900612.

54. Interview with Iyad Kheirbek of [New York City Department of Mental Health and Hygiene], January 2015.

55. "2014 West Africa Ebola Response—OpenStreetMap Wiki," accessed January 25, 2019, https://wiki.openstreetmap.org/wiki/2014_West_Africa_Ebola_Response.

56. "Ushahidi," accessed January 25, 2019, https://www.ushahidi.com/.

57. Ida Norheim-Hagtun and Patrick Meier, "Crowdsourcing for Crisis Mapping in Haiti," *Innovations: Technology, Governance, Globalization* 5, no. 4 (2010): 81–89.

58. Jessica Ramirez, "'Ushahidi' Technology Saves Lives in Haiti and Chile," *Newsweek*, March 3, 2010, https://www.newsweek.com/ushahidi-technology-saves-lives-haiti-and-chile-210262.

59. Norheim-Hagtun and Meier, "Crowdsourcing for Crisis Mapping in Haiti."

60. Patrick Meier, "Crowdsourcing the Evaluation of Post-Sandy Building Damage Using Aerial Imagery," *IRevolutions* (blog), November 1, 2012, https://irevolutions.org/2012/11/01/crowdsourcing-sandy-building-damage/.

61. Introduction to Data Science (IDA), "About," IDS webpage, https://www.idsucla.org/about.

62. Ron Eglash, Juan E. Gilbert, and Ellen Foster, "Toward Culturally Responsive Computing Education," *Commun. ACM* 56, no. 7 (July 2013): 33–36, https://doi.org/10.1145/2483852.2483864; Amelia McNamara and Mark Hansen, "Teaching Data Science to Teenagers," in *Proceedings of the Ninth International Conference on Teaching Statistics*, 2014.

63. Nicole Lazar and Christine Franklin, "The Big Picture: Preparing Students for a Data-Centric World," *Chance* 28, no. 4 (2015): 43–45.

Chapter 3

1. Steven Levy, *Hackers: Heroes of the Computer Revolution*, vol. 14 (Garden City, NY: Anchor Press/Doubleday, 1984).

2. Patrick Greenfield, "The Cambridge Analytica Files: The Story so Far," *Guardian*, March 25, 2018, https://www.theguardian.com/news/2018/mar/26/the-cambridge-analytica-files-the-story-so-far.

3. Annabel Latham, "Cambridge Analytica Scandal: Legitimate Researchers Using Facebook Data Could Be Collateral Damage," *The Conversation*, accessed July 6, 2019, http://theconversation.com/cambridge-analytica-scandal-legitimate-researchers-using-facebook-data-could-be-collateral-damage-93600.

4. "Reporter Shows the Links between the Men behind Brexit and the Trump Campaign," NPR.org, accessed July 6, 2019, https://www.npr.org/2018/07/19/630443485/reporter-shows-the-links-between-the-men-behind-brexit-and-the-trump-campaign.

5. Matthew Rosenberg, "Academic behind Cambridge Analytica Data Mining Sues Facebook for Defamation," *New York Times*, March 15, 2019, https://www.nytimes.com/2019/03/15/technology/aleksandr-kogan-facebook-cambridge-analytica.html.

6. Matthew Zook et al., "Ten Simple Rules for Responsible Big Data Research," *PLOS Computational Biology* 13, no. 3 (March 30, 2017): e1005399, https://doi.org/10.1371/journal.pcbi.1005399.

7. One notable way of alienating entire communities is by crime mapping: creating hot-spot analyses to pinpoint where crimes take place is to mark clusters as criminal neighborhoods. Careful ethical understanding and review of data is in order to minimize the stigmatization of entire groups of residents.

8. Todd Haselton, "Google Still Keeps a List of Everything You Ever Bought Using Gmail, Even If You Delete All Your Emails," CNBC, July 5, 2019, https://www.cnbc.com/2019/07/05/google-gmail-purchase-history-cant-be-deleted.html.

9. "Identity Resolution Service & Data Onboarding," LiveRamp, accessed July 13, 2019, https://liveramp.com/.

10. During the same year, 2008, we also find one of the first examples of unethical practices of social media use. Harvard provided researchers with all the Facebook handles of their freshman class, which were then scraped and cross-referenced with other data. Although the data collected from this project removed identifiable information such as names, emails, phone, and numbers, within hours of posting the data other researchers figured out who the participants were.

11. Mingzhe Tang and N. Edward Coulson, "The Impact of China's Housing Provident Fund on Homeownership, Housing Consumption and Housing Investment," *Regional Science and Urban Economics* 63 (2017): 25–37.

12. Edward Glaeser et al., "A Real Estate Boom with Chinese Characteristics," *Journal of Economic Perspectives* 31, no. 1 (2017): 93–116.

13. Michael Batty, "At the Crossroads of Urban Growth," *Environment and Planning B: Planning and Design* 41 (2014): 951–953; Jing Wu, Joseph Gyourko, and Yongheng Deng, "Evaluating

the Risk of Chinese Housing Markets: What We Know and What We Need to Know," *China Economic Review* 39 (July 1, 2016): 91–114, https://doi.org/10.1016/j.chieco.2016.03.008; Wade Shepard, *Ghost Cities of China: The Story of Cities without People in the World's Most Populated Country* (London: Zed Books Ltd., 2015); Christian Sorace and William Hurst, "China's Phantom Urbanisation and the Pathology of Ghost Cities," *Journal of Contemporary Asia* 46, no. 2 (2016): 304–322.

14. Shepard, *Ghost Cities of China*.

15. Guanghua Chi et al., "Ghost Cities Analysis Based on Positioning Data in China," ArXiv:1510.08505 [cs.SI], October 28, 2015, http://arxiv.org/abs/1510.08505.

16. Xiaobin Jin et al., "Evaluating Cities' Vitality and Identifying Ghost Cities in China with Emerging Geographical Data," *Cities* 63 (March 2017): 98–109, https://doi.org/10.1016/j.cities.2017.01.002; Yongling Yao et al., "House Vacancy at Urban Areas in China with Nocturnal Light Data of DMSP-OLS," *Proceedings of 2011 IEEE International Conference on Spatial Data Mining and Geographical Knowledge Services*, June 29–Jult, 2011, https://doi.org/10.1109/ICSDM.2011.5969087; Chi et al., "Ghost Cities Analysis Based on Positioning Data in China."

17. Predictive data models often use statistical methods. If you are interested in developing predictive models, a statistics class is a great start. Or having a data scientist or statistician on your team. A detailed explanation of the models has been published recently in Sarah Williams, Wenfei Xu, Shin Bin Tan, Michael J. Foster, and Changping Chen, "Ghost Cities of China: Identifying Urban Vacancy through Social Media Data." *Cities* 94 (2019): 275–285.

18. It is important to note here that while this description of the project makes it seem as though it followed a strictly linear trajectory, the process was much more like a wavy line with lots of discovery along the way. For example, learning to download the data to create a complete data set was not easy. And we didn't know exactly what data we might use, although we had a hypothesis. Finding out that certain data sets do not work is also important to work like this.

19. Sherwin Rosen, "Wage-Based Indexes of Urban, Quality of Life," *Current Issues in Urban Economics*, 1979, 74–104.

20. Edward L. Glaeser, Jed Kolko, and Albert Saiz, "Consumer City," *Journal of Economic Geography* 1, no. 1 (2001): 27–50, https://doi.org/10.1093/jeg/1.1.27.

21. Siqi Zheng and Matthew E. Kahn, "Land and Residential Property Markets in a Booming Economy: New Evidence from Beijing," *Journal of Urban Economics* 63, no. 2 (March 1, 2008): 743–757, https://doi.org/10.1016/j.jue.2007.04.010; Siqi Zheng, Yuming Fu, and Hongyu Liu, "Demand for Urban Quality of Living in China: Evolution in Compensating Land-Rent and Wage-Rate Differentials," *Journal of Real Estate Finance and Economics* 38, no. 3 (April 1, 2009): 194–213, https://doi.org/10.1007/s11146-008-9152-0; W. Ding, S. Zheng, and X. Guo, "Value of Access to Jobs and Amenities: Evidence from New Residential Properties in Beijing," *Tsinghua Science and Technology* 15, no. 5 (October 2010): 595–603, https://doi.org/10.1016/S1007-0214(10)70106-1; Daisy J. Huang, Charles K. Leung, and Baozhi Qu, "Do Bank Loans and Local Amenities Explain Chinese Urban House Prices?," *China Economic Review* 34 (July 1, 2015): 19–38, https://doi.org/10.1016/j.chieco.2015.03.002.

22. It should be noted that data scraped from Chinese social media sites often has geopositional errors imposed by the Chinese government to make it more difficult to copy the data, but there are many algorithms online that allow one to correct for these errors. We had to apply these corrections to all the data we used.

23. Wei Luo and Fahui Wang, "Measures of Spatial Accessibility to Health Care in a GIS Environment: Synthesis and a Case Study in the Chicago Region," *Environment and Planning B: Planning and Design* 30, no. 6 (December 1, 2003): 865–884, https://doi.org/10.1068/b29120; Mark F Guagliardo et al., "Physician Accessibility: An Urban Case Study of Pediatric Providers," *Health & Place* 10, no. 3 (September 1, 2004): 273–283, https://doi.org/10.1016/j.healthplace.2003.01.001; Mark F. Guagliardo, "Spatial Accessibility of Primary Care: Concepts, Methods and Challenges," *International Journal of Health Geographics* 3, no. 1 (February 26, 2004): 3, https://doi.org/10.1186/1476-072X-3-3.

24. David M. Levinson, "Accessibility and the Journey to Work," *Journal of Transport Geography* 6, no. 1 (March 1, 1998): 11–21, https://doi.org/10.1016/S0966-6923(97)00036-7; Mei-Po Kwan, "Gender, the Home-Work Link, and Space-Time Patterns of Nonemployment Activities," *Economic Geography* 75, no. 4 (October 1, 1999): 370–394, https://doi.org/10.1111/j.1944-8287.1999.tb00126.x; Selima Sultana, "Job/Housing Imbalance and Commuting Time in the Atlanta Metropolitan Area: Exploration of Causes of Longer Commuting Time," *Urban Geography* 23, no. 8 (December 1, 2002): 728–749, https://doi.org/10.2747/0272-3638.23.8.728; Fahui Wang and W. William Minor, "Where the Jobs Are: Employment Access and Crime Patterns in Cleveland," *Annals of the Association of American Geographers* 92, no. 3 (2002): 435–450, https://doi.org/10.1111/1467-8306.00298; Paul Waddell and Gudmundur F. Ulfarsson, *Accessibility and Agglomeration: Discrete-Choice Models of Employment Location by Industry Sector*, vol. 63, 82nd Annual Meeting of the Transportation Research Board, Washington, DC, 2003; Mark W. Horner, "Spatial Dimensions of Urban Commuting: A Review of Major Issues and Their Implications for Future Geographic Research," *Professional Geographer* 56, no. 2 (May 1, 2004): 160–173, https://doi.org/10.1111/j.0033-0124.2004.05602002.x; Daniel Baldwin Hess, "Access to Employment for Adults in Poverty in the Buffalo-Niagara Region," *Urban Studies* 42, no. 7 (June 1, 2005): 1177–1200, https://doi.org/10.1080/00420980500121384.

25. Susan L. Handy, "Regional Versus Local Accessibility: Neo-Traditional Development and Its Implications for Non-Work Travel on JSTOR," *Built Environment* 18, no. 4 (1992): 253–267.

26. Chandra Bhat, Jay Carini, and Rajul Misra, "Modeling the Generation and Organization of Household Activity Stops," *Transportation Research Record: Journal of the Transportation Research Board* 1676 (January 1, 1999): 153–161, https://doi.org/10.3141/1676-19.

27. Walter G. Hansen, "How Accessibility Shapes Land Use," *Journal of the American Institute of Planners* 25, no. 2 (May 1, 1959): 73–76, https://doi.org/10.1080/01944365908978307.

28. We called the results "underutilized" because the cities often were not truly vacant. We felt it was important to clarify the results and be precise, since underutilized land could be readied for development. The categories included (1) under construction; (2) halted construction; (3) vacant or built but no one moved in; and (4) older communist housing (empty or partially occupied).

29. For example, knowing that you residential location got a low amenities score because the nearest school was far away could tell the urban planners in the city to build a new school.

30. Interview with Yixue Jiao, October 12, 2016.

31. Brent Hecht and Monica Stephens, "A Tale of Cities: Urban Biases in Volunteered Geographic Information," *ICWSM* 14 (2014): 197–205.

32. Luc Anselin and Sarah Williams. "Digital Neighborhoods." *Journal of Urbanism: International Research on Placemaking and Urban Sustainability* 9, no. 4 (October 1, 2016): 305–328. https://doi.org/10.1080/17549175.2015.1080752.

33. Melissa Parrish et al., "Location-Based Social Networks: A Hint of Mobile Engagement Emerges," *Forrester Research*, 2010; Derek Ruths and Jürgen Pfeffer, "Social Media for Large Studies of Behavior," *Science* 346, no. 6213 (November 28, 2014): 1063–1064, https://doi.org/10.1126/science.346.6213.1063; Maeve Duggan and Joanna Brenner, *The Demographics of Social Media Users*, 2012, vol. 14 (Pew Research Center's Internet & American Life Project Washington, DC, 2013).

34. Anselin and Williams, "Digital Neighborhoods."

35. Francis Harvey, "To Volunteer or to Contribute Locational Information? Towards Truth in Labeling for Crowdsourced Geographic Information," in *Crowdsourcing Geographic Knowledge: Volunteered Geographic Information (VGI) in Theory and Practice*, ed. Daniel Sui, Sarah Elwood, and Michael Goodchild (Dordrecht: Springer Netherlands, 2013), 31–42, https://doi.org/10.1007/978-94-007-4587-2_3.

36. Elizabeth Currid and Sarah Williams, "The Geography of Buzz: Art, Culture and the Social Milieu in Los Angeles and New York," *Journal of Economic Geography* 10, no. 3 (2009): 423–451.

37. Melena Ryzik, "'The Geography of Buzz,' a Study on the Urban Influence of Culture," *New York Times*, April 6, 2009, https://www.nytimes.com/2009/04/07/arts/design/07buzz.html.

38. Kevin Lynch, *The Images of the City* (Cambridge, MA: MIT Press), 1960.

39. Kevin Lynch, *Some Major Problems, Boston: MIT Libraries*, 1960, Perceptual Map, 1960, MC 208, Box 6, MIT Dome, https://dome.mit.edu/handle/1721.3/36515.

40. Sarah Williams, "Here Now! Social Media and the Psychological City," in *Inscribing a Square—Urban Data as Public Space* (New York: Springer, 2012).

41. Guy Lansley and Paul A. Longley, "The Geography of Twitter Topics in London," *Computers, Environment and Urban Systems* 58 (2016): 85–96; Bernd Resch et al., "Citizen-Centric Urban Planning through Extracting Emotion Information from Twitter in an Interdisciplinary Space-Time-Linguistics Algorithm," *Urban Planning* 1, no. 2 (2016): 114–127, https://doi.org/10.17645/up.v1i2.617.

42. Anna Kovacs-Gyori et al., "#London2012: Towards Citizen-Contributed Urban Planning Through Sentiment Analysis of Twitter Data," *Urban Planning* 3, no. 1 (2018): 75–99, https://doi.org/10.17645/up.v3i1.1287; Ayelet Gal-Tzur et al., "The Potential of Social Media in Delivering Transport Policy Goals," *Transport Policy* 32 (2014): 115–123, https://doi.org/10.1016/j.tranpol.2014.01.007.

43. Gary W. Evans, *Environmental Stress* (Cambridge: Cambridge University Press, 1984).

44. Luca Maria Aiello, Rossano Schifanella, Danielle Quercia, and Francesco Aietta, "Chatty Maps: Constructing Sound Maps of Urban Areas from Social Media Data," *Royal Society Open Science* 3, no. 3 (March 1, 2016): 150690, https://doi.org/10.1098/rsos.150690.

45. Ibid.

46. Daniele Quercia et al., "Smelly Maps: The Digital Life of Urban Smellscapes," ArXiv:1505.06851 [Cs.SI], May 26, 2015, http://arxiv.org/abs/1505.06851.

47. Aiello et al., "Chatty Maps."

48. Anselin and Williams, "Digital Neighborhoods."

49. Justin Cranshaw et al., "The Livehoods Project: Utilizing Social Media to Understand the Dynamics of a City," *Proceedings of the Sixth International AAAI Conference on Weblogs and Social Media*, 2012.

50. T. W. Grein et al., "Rumors of Disease in the Global Village: Outbreak Verification," *Emerging Infectious Diseases* 6, no. 2 (2000): 97–102; David L Heymann and Guénaël R Rodier, "Hot Spots in a Wired World: WHO Surveillance of Emerging and Re-Emerging Infectious Diseases," *Lancet Infectious Diseases* 1, no. 5 (December 1, 2001): 345–353, https://doi.org/10.1016/S1473-3099(01)00148-7; Nkuchia M. M'ikanatha et al., "Use of the Internet to Enhance Infectious Disease Surveillance and Outbreak Investigation," *Biosecurity and Bioterrorism: Biodefense Strategy, Practice, and Science* 4, no. 3 (September 1, 2006): 293–300, https://doi.org/10.1089/bsp.2006.4.293.

51. John S. Brownstein et al., "Surveillance Sans Frontières: Internet-Based Emerging Infectious Disease Intelligence and the HealthMap Project," *PLOS Medicine* 5, no. 7 (July 8, 2008): e151, https://doi.org/10.1371/journal.pmed.0050151.

52. Rumi Chunara, Jason R. Andrews, and John S. Brownstein, "Social and News Media Enable Estimation of Epidemiological Patterns Early in the 2010 Haitian Cholera Outbreak," *American Journal of Tropical Medicine and Hygiene* 86, no. 1 (January 1, 2012): 39–45, https://doi.org/10.4269/ajtmh.2012.11-0597.

53. Taha A. Kass-Hout and Hend Alhinnawi, "Social Media in Public Health," *British Medical Bulletin* 108, no. 1 (2013): 5–24.

54. "Humanitarian Tracker," Humanitarian Tracker, October 15, 2018, https://www.humanitariantracker.org.

55. Domenico Giannone, Lucrezia Reichlin, and David Small, "Nowcasting: The Real-Time Informational Content of Macroeconomic Data," *Journal of Monetary Economics* 55, no. 4 (May 2008): 665–676, https://doi.org/10.1016/j.jmoneco.2008.05.010.

56. Marta Banbura et al., "Now-Casting and the Real-Time Data Flow," *Handbook of Economic Forecasting* 2, no. Part A (2013): 195–237.

57. Ali Alessa and Miad Faezipour, "A Review of Influenza Detection and Prediction through Social Networking Sites," *Theoretical Biology & Medical Modelling* 15 (February 1, 2018), https://doi.org/10.1186/s12976-017-0074-5.

58. Jeremy Ginsberg et al., "Detecting Influenza Epidemics Using Search Engine Query Data," *Nature* 457, no. 7232 (February 2009): 1012–1014, https://doi.org/10.1038/nature07634.

59. Ginsberg et al., "Detecting Influenza Epidemics."

60. Miguel Helft, "Google Uses Web Searches to Track Flu's Spread," *New York Times*, November 11, 2008, https://www.nytimes.com/2008/11/12/technology/internet/12flu.html; "The Next Chapter for Flu Trends," *Google AI Blog* (blog), October 15, 2018, http://ai.googleblog.com/2015/08/the-next-chapter-for-flu-trends.html.

61. Samantha Cook et al., "Assessing Google Flu Trends Performance in the United States during the 2009 Influenza Virus A (H1N1) Pandemic," *PLOS ONE* 6, no. 8 (August 19, 2011): e23610, https://doi.org/10.1371/journal.pone.0023610.

62. Donald R. Olson et al., "Reassessing Google Flu Trends Data for Detection of Seasonal and Pandemic Influenza: A Comparative Epidemiological Study at Three Geographic Scales," *PLOS Computational Biology* 9, no. 10 (October 17, 2013): e1003256, https://doi.org/10.1371/journal.pcbi.1003256.

63. David Lazer et al., "The Parable of Google Flu: Traps in Big Data Analysis," *Science* 343, no. 6176 (March 14, 2014): 1203–5, https://doi.org/10.1126/science.1248506.

64. Joseph Jeffrey Klembczyk et al., "Google Flu Trends Spatial Variability Validated Against Emergency Department Influenza-Related Visits," *Journal of Medical Internet Research* 18, no. 6 (2016): e175, https://doi.org/10.2196/jmir.5585.

65. "What We Can Learn from the Epic Failure of Google Flu Trends," *Wired*, October 17, 2018, https://www.wired.com/2015/10/can-learn-epic-failure-google-flu-trends/.

66. Laura Bliss, "The Imperfect Science of Mapping the Flu," CityLab, January 20, 2018, https://www.citylab.com/design/2018/01/the-imperfect-science-of-mapping-the-flu/551387

67. Ryan Burns, "Moments of Closure in the Knowledge Politics of Digital Humanitarianism," *Geoforum* 53 (May 1, 2014): 51–62, https://doi.org/10.1016/j.geoforum.2014.02.002.

68. Sarah Vieweg et al., "Microblogging during Two Natural Hazards Events: What Twitter May Contribute to Situational Awareness," in *Proceedings of the SIGCHI Conference on Human Factors in Computing Systems* (ACM, 2010), 1079–1088; Kate Starbird et al., "Chatter on the Red: What Hazards Threat Reveals about the Social Life of Microblogged Information," in *Proceedings of the 2010 ACM Conference on Computer Supported Cooperative Work* (ACM, 2010), 241–250; Kate Starbird and Leysia Palen, *Pass It On?: Retweeting in Mass Emergency* (International Community on Information Systems for Crisis Response and Management, 2010); Paul S. Earle, Daniel C. Bowden, and Michelle Guy, "Twitter Earthquake Detection: Earthquake Monitoring in a Social World," *Annals of Geophysics* 54, no. 6 (January 14, 2012), https://doi.org/10.4401/ag-5364.

69. Stephanie Busari, "Tweeting the Terror: How Social Media Reacted to Mumbai," CNN.com, 2008, https://www.cnn.com/2008/WORLD/asiapcf/11/27/mumbai.twitter/.

70. J. Rogstadius et al., "CrisisTracker: Crowdsourced Social Media Curation for Disaster Awareness," *IBM Journal of Research and Development* 57, no. 5 (September 2013): 4:1–4:13, https://doi.org/10.1147/JRD.2013.2260692.

71. Nezih Altay and Melissa Labonte, "Challenges in Humanitarian Information Management and Exchange: Evidence from Haiti," *Disasters* 38, no. s1 (2014): S50–72, https://doi.org/10.1111/disa.12052.

72. Kate Crawford and Megan Finn, "The Limits of Crisis Data: Analytical and Ethical Challenges of Using Social and Mobile Data to Understand Disasters," *GeoJournal* 80, no. 4 (August 1, 2015): 491–502, https://doi.org/10.1007/s10708-014-9597-z.

73. Altay and Labonte, "Challenges in Humanitarian Information Management and Exchange"; Nathan Morrow et al., "Independent Evaluation of the Ushahidi Haiti Project," 2011, https://www.researchgate.net/publication/265059793_Ushahidi_Haiti_Project_Evaluation_ Independent_Evaluation_of_the_Ushahidi_Haiti_Project.

74. Charles Duhigg, "How Companies Learn Your Secrets," *New York Times Magazine*, February 16, 2012, https://www.nytimes.com/2012/02/19/magazine/shopping-habits.html.

Chapter 4

1. "James Madison to W. T. Barry, August 4, 1822," image, Library of Congress, Washington, DC, accessed January 26, 2019, https://www.loc.gov/resource/mjm.20_0155_0159/?st=list.

2. William Playfair, *Commercial and Political Atlas and Statistical Breviary (Original Version Was Published in 1786)*, 2.

3. Miles A. Kimball, "London through Rose-Colored Graphics: Visual Rhetoric and Information Graphic Design in Charles Booth's Maps of London Poverty," *Journal of Technical Writing and Communication* 36, no. 4 (October 2006): 353–81, https://doi.org/10.2190/ K561-40P2-5422-PTG2. Playfair was a curious fellow; although he put his mind to many great inventions of his time, he ended up in the poorhouse as a result of his involvement in numerous dubious financial schemes. His character often comes into question in the discussion of his visualizations; he was so good at creating them as a means of persuasion that one wonders if they were an artifice of his many schemes.

4. Bekkers and Moody, "Visual Events and Electronic Government."

5. Manuel Castells, *Communication Power* (Oxford: Oxford University Press, 2013); Victor Bekkers and Rebecca Moody, "Visual Events and Electronic Government: What Do Pictures Mean in Digital Government for Citizen Relations?," *Government Information Quarterly* 28, no. 4 (October 1, 2011): 457–465, https://doi.org/10.1016/j.giq.2010.10.006.

6. "'Stowage of the British Slave Ship "Brookes" under the Regulated Slave Trade, Act of 1788,'" The Abolition Seminar, accessed July 19, 2019, https://www.abolitionseminar.org/brooks/.

7. Andrew Milne-Skinner, "Liverpool's Slavery Museum: A Blessing or a Blight?'" *Racism, Slavery, and Literature* (Frankfurt/M.: Peter Lang 2010), 2010, 183–195.

8. Eyal Weizman, "Introduction: Forensis," *Forensis: The Architecture of Public Truth*, 2014, 9–32.

9. "The Brookes—Visualising the Transatlantic Slave Trade," accessed July 19, 2019, https:// www.history.ac.uk/1807commemorated/exhibitions/museums/brookes.html. In this respect the image is a symbol of control; it dictates a particular way of viewing and remembering the slave trade. Significantly, it offers only one perspective, that of the abolitionist witnessing cruelty—the perspective of those who are suffering the cruelty is predominantly absent.

10. Sarah Williams, Elizabeth Marcello, and Jacqueline M. Klopp, "Toward Open Source Kenya: Creating and Sharing a GIS Database of Nairobi," *Annals of the Association of American Geographers* 104, no. 1 (January 2, 2014): 114–130, https://doi.org/10.1080/00045608.2013.846157.

11. Perhaps even more interestingly, a recent Japan International Cooperation Agency (JICA) master plan uses this land-use map. The agency couldn't get access to the original land-use map made for the city, which was used to create this open source map.

12. It's important to note that maps are part of Nairobi's cultural references, where education is largely based on the British system, and includes work in geography.

13. Sarah Williams et al., "Digital Matatus: Using Mobile Technology to Visualize Informality," *Proceedings ACSA 103rd Annual Meeting: The Expanding Periphery and the Migrating Center*, 2015.

14. Gilbert Koech and Chrispinus Wekesa, "Kenya: Nairobi Transit Map Launched," *Star*, January 29, 2014, https://allafrica.com/stories/201401290479.html.

15. @ma3route, "Ma3Route Is a Mobile/Web/SMS Platform That Crowd-Sources for Transport Data and Provides Users with Information on Traffic, Matatu Directions and Driving Reports," accessed August 1, 2019, http://www.ma3route.com/.

16. Gauff Consultants, *Nairobi / Proposed MRTS Commuter Rail Network*, 2014.

17. Jeremy W. Crampton, "Maps as Social Constructions: Power, Communication and Visualization," *Progress in Human Geography* 25, no. 2 (2001): 235–252.

18. David Harvey, "On the History and Present Condition of Geography: An Historical Materialist Manifesto," *Professional Geographer* 36, no. 1 (February 1, 1984): 1–11, https://doi.org/10.1111/j.0033-0124.1984.00001.x.

19. J. Brian Harley, "Maps, Knowledge, and Power," *Geographic Thought: A Praxis Perspective*, 2009, 129–148.

20. Ibid.

21. Laura Kurgan, *Close Up at a Distance: Mapping, Technology, and Politics* (Cambridge, MA: MIT Press, 2013).

22. Kurgan, *Close Up at a Distance*.

23. Michelle Alexander, *The New Jim Crow: Mass Incarceration in the Age of Colorblindness* (New York: The New Press, 2012).

24. DataMade, "Chicago's Million Dollar Blocks," January 7, 2019, https://chicagosmilliondollarblocks.com/.

25. Laura Kurgan, director of the Spatial Information Design lab and the principal investigator (PI) for the grant, brought this diverse group together.

26. Australian law Reform Comission, "How Does Justice Reinvestment Work?" November 1, 2018. https://www.alrc.gov.au/publication/pathways-to-justice-inquiry-into-the-incarceration-rate-of-aboriginal-and-torres-strait-islander-peoples-alrc-report-133/4-justice-reinvestment/how-does-justice-reinvestment-work/.

27. Sir Patrick Geddes, *Town Planning in Kapurthala: A Report to the HH The Maharaja of Kapurthala*, 1917.

28. Christine M. Boyer, *Dreaming the Rational City: The Myth of American City Planning* (Cambridge, MA: MIT Press, 1986). Geddes developed a project called Camera Obscura, which asked users to view the real-time interplay of cities through an observation tower. He thought that observing the city at a distance in this way would allow us to better understand how people have an effect on one another. While a somewhat abstract project, it is clear the Geddes struggled with the notion of a one-dimensional map—and this was his attempt to get planners to see the city from a different vantage point. I find Geddes's fascination with trying to find new ways of seeing places inspiring and at the same time strange.

29. Helen Meller, *Patrick Geddes: Social Evolutionist and City Planner* (London: Routledge, 2005).

30. Ibid.

31. Volker M. Welter and Iain Boyd Whyte, *Biopolis: Patrick Geddes and the City of Life* (Cambridge, MA: MIT Press, 2002).

32. Sarah Williams et al., "City Digits: Local Lotto: Developing Youth Data Literacy by Investigating the Lottery," *Journal of Digital and Media Literacy*, December 15. 2014, http://civicdatadesignlab.mit.edu/files/youth-digital-literacy.pdf.

33. Ibid.

34. Ibid.

35. Sarah Williams, "Data Visualizations Break Down Knowledge Barriers in Public Engagement," in *The Civic Media Reader*, ed. Eric Gordon and Paul Mihailidis (Cambridge, MA: MIT Press, 2016).

36. Ford Fessenden and Josh Keller, "How Minorities Have Fared in States with Affirmative Action Bans," January 7, 2019, https://www.nytimes.com/interactive/2013/06/24/us/affirmative-action-bans.html.

37. Wilson Huhn, *Schuette v. Coalition to Defend Affirmative Action*, US Supreme Court, No. 572 U.S. 20142013, 2013.

38. Ibid.

39. "Do Affirmative Action Bans Hurt Minority Students?," IVN.us, April 29, 2014, https://ivn.us/2014/04/29/affirmative-action-bans-hurt-minority-students-states-show-mixed-results/.

40. "Correcting the *New York Times* on College 'Enrollment Gaps,'" AEI, June 25, 2013, http://www.aei.org/publication/correcting-the-new-york-times-on-college-enrollment-gaps/.

41. "*New York Times* Excludes Asian-Americans from Affirmative Action Study," January 7, 2019, https://dailycaller.com/2014/04/23/new-york-times-excludes-asian-americans-from-affirmative-action-study/.

42. Jamiles Lartey, "US Police Killings Undercounted by Half, Study Using Guardian Data Finds," *Guardian*, October 11, 2017, https://www.theguardian.com/us-news/2017/oct/11/police-killings-counted-harvard-study.

43. Jon Swaine, "Police Will Be Required to Report Officer-Involved Deaths under New US System," *Guardian*, August 8, 2016, https://www.theguardian.com/us-news/2016/aug/08/police-officer-related-deaths-department-of-justice.

44. Jeanne Bourgault, "How the Global Open Data Movement Is Transforming Journalism," *Wired*, 2013, https://www.wired.com/insights/2013/05/how-the-global-open-data-movement-is-transforming-journalism/; Brett Goldstein and Lauren Dyson, eds., *Beyond Transparency: Open Data and the Future of Civic Innovation* (San Francisco: Code for America Press, 2013).

45. Fagan notes in his study that stop-and-frisk policies stemmed from a form of policing that was focused on policing areas with suspicious for criminal behavior. Communities that had crime "hot spots" or those that appeared to be in disrepair (e.g., with broken windows) were all suspect. These ideologies were transferred from physical space to people who were suspicious, giving the police the right to perform a search for suspicion.

46. Jeffrey Fagan and Amanda Geller, "Following the Script: Narratives of Suspicion in Terry Stops in Street Policing," SSRN Scholarly Paper (Rochester, NY: Social Science Research Network, August 1, 2014), https://papers.ssrn.com/abstract=2485375.

47. "Case Study: *Floyd v NYC*," Catalysts for Collaboration, accessed January 10, 2019, https://catalystsforcollaboration.org/casestudy/nycfloyd.html.

48. Ibid.

49. "NYCLU_2011_Stop-and-Frisk_Report.Pdf," accessed January 10, 2019, https://www.nyclu.org/sites/default/files/publications/NYCLU_2011_Stop-and-Frisk_Report.pdf.

50. Jason Oberholtzer, "Stop-And-Frisk by the Numbers," *Forbes*, 2012, https://www.forbes.com/sites/jasonoberholtzer/2012/07/17/stop-and-frisk-by-the-numbers/#5a02ed656703.

51 "All the Stops," *BKLYNR*, accessed January 10, 2019, https://www.bklynr.com/all-the-stops/.

52. WNYC map, from "NYPD Stop-and-Frisk Data," accessed January 10, 2019, https://people.cs.uct.ac.za/~mkuttel/VISProjects2016/4AllieEtAl/index.html.

53. "List of Physical Visualizations," accessed January 27, 2019, http://dataphys.org/list/stop-and-frisk-physical-data-filtering/.

54. Daniel Denvir, "The Key Ingredient in Stop and Frisk Reform: Open Data," CityLab, August 25, 2015, http://www.citylab.com/crime/2015/08/the-missing-ingredient-in-stop-and-frisk-accountability-open-data/402026/; Sam Blum, "New Data Visualizations Reveal the Racial Disparity of Stop and Frisk," *Brooklyn Magazine*, June 23, 2015, http://www.bkmag.com/2015/06/23/new-data-visualizations-reveal-the-racial-disparity-of-stop-and-frisk/; Tracey Meares, "Programming Errors: Understanding the Constitutionality of Stop and Frisk as a Program, Not an Incident," 2014, https://doi.org/10.13140/2.1.4252.6404.

55. "Landmark Decision: Judge Rules NYPD Stop and Frisk Practices Unconstitutional, Racially Discriminatory," Center for Constitutional Rights, accessed July 21, 2019, https://ccrjustice.org/node/1269.

56. Barack Obama, "Transparency and Open Government," Memorandum for the Heads of Executive Departments and Agencies, January 21, 2009, https://obamawhitehouse.archives.gov/the-press-office/transparency-and-open-government.

Chapter 5

1. The Economist, "Off the Map," *Economist*, November 13, 2014, https://www.economist.com/news/international/21632520-rich-countries-are-deluged-data-developing-ones-are-suffering-drought.

2. Umar Serajuddin et al., *Data Deprivation: Another Deprivation to End* (Washington, DC: World Bank, 2015).

3. Ibid.

4. Data Revolution Group, "A World That Counts," 2014, http://www.undatarevolution.org/wp-content/uploads/2014/11/A-World-That-Counts.pdf.

5. danah boyd and Kate Crawford, "Critical Questions for Big Data: Provocations for a Cultural, Technological, and Scholarly Phenomenon," *Information, Communication & Society* 15, no. 5 (2012): 662–679.

6. Sean O'Kane, "Tesla and Waymo Are Taking Wildly Different Paths to Creating Self-Driving Cars," *Verge*, April 19, 2018, https://www.theverge.com/transportation/2018/4/19/17204044/tesla-waymo-self-driving-car-data-simulation.

7. Sarah Williams, "Who Owns the City of the Future?" *Cooper Hewitt Design Journal* (Winter 2018): 8–10.

8. Jim Thatcher, David O'Sullivan, Dillon Mahmoudi, "Data Colonialism through Accumulation by Dispossession: New Metaphors for Daily Data," *Environment and Planning D: Society and Space* 34, no. 6 (2016): 990–1006.

9. Nick Couldry, *The Costs of Connection: How Data Is Colonizing Human Life and Appropriating It for Capitalism* (Stanford, CA: Stanford University Press, 2019).

10. Greg Elmer, Ganaele Langlois, and Joanna Redden, *Compromised Data: From Social Media to Big Data* (New York: Bloomsbury Publishing, 2015).

11. Jo Bates, "The Strategic Importance of Information Policy for the Contemporary Neoliberal State: The Case of Open Government Data in the United Kingdom," *Government Information Quarterly* 31 (2014): 388–395.

12. Thomas L. McPhail and Everett M. Rogers, *Electronic Colonialism: The Future of International Broadcasting and Communication* (Beverly Hills, CA: Sage, 1981).

13. Linnet Taylor, "The Ethics of Big Data as a Public Good: Which Public? Whose Good?," *Phil. Trans. R. Soc. A* 374, no. 2083 (2016): 20160126.

14. Hal R. Varian, *Microeconomic Analysis, Third Edition*, 3rd edition (New York: Norton, 1992).

15. Rafael Melgarejo-Heredia, Leslie Carr, and Susan Halford, "The Public Web and the Public Good," in *Proceedings of the 8th ACM Conference on Web Science*, WebSci '16 (New York: ACM, 2016), 330–332, https://doi.org/10.1145/2908131.2908181.

16. Joseph E. Stiglitz, *Knowledge as a Global Public Good* (New York: Oxford University Press, 1999), https://www.oxfordscholarship.com/view/10.1093/0195130529.001.0001/acprof-9780195130522-chapter-15; Hal R. Varian, *Markets for Information Goods*, vol. 99 (Citeseer, 1999).

17. Linnet Taylor, "The Ethics of Big Data as a Public Good: Which Public? Whose Good?," SSRN Scholarly Paper (Rochester, NY: Social Science Research Network, August 9, 2016), https://papers.ssrn.com/abstract=2820580.

18. Nadezhda Purtova, "Illusion of Personal Data as No One's Property," SSRN Scholarly Paper (Rochester, NY: Social Science Research Network, October 29, 2013), https://papers.ssrn.com/abstract=2346693.

19. Mike Maciag, "How Driverless Cars Could Be a Big Problem for Cities," Governing, 2017, https://www.governing.com/topics/finance/gov-cities-traffic-parking-revenue-driverless-cars.html.

20. Gideon Mann, "Private Data and the Public Good," May 17, 2016, https://medium.com/@gideonmann/private-data-and-the-public-good-9c94c656ff28. https://medium.com/@gideonmann/private-data-and-the-public-good-9c94c656ff28Mann.

21. Michael Barbaro and Tom Zeller Jr, "A Face Is Exposed for AOL Searcher No. 4417749," *New York Times*, August 9, 2006, https://www.nytimes.com/2006/08/09/technology/09aol.html.

22. Katie Hafner, "Researchers Yearn to Use AOL Logs, but They Hesitate," *New York Times*, August 23, 2006, https://www.nytimes.com/2006/08/23/technology/23search.html.

23. Sean M. McDonald, "Ebola: A Big Data Disaster-Privacy, Property, and the Law of Disaster Experimentation," *Centre for Internet and Society*, no. 2016.01 (2016).

24. Organisation for Economic Cooperation and Development (OECD), "OECD Guidelines on the Protection of Privacy and Transborder Flows of Personal Data," September 23, 1980, http://www.oecd.org/internet/ieconomy/oecdguidelinesontheprotectionofprivacyandtransborderflowsofpersonaldata.htm.

25. European Union, "2018 Reform of EU Data Protection Rules," 2018, https://ec.europa.eu/commission/priorities/justice-and-fundamental-rights/data-protection/2018-reform-eu-data-protection-rules_en.

26. Natasha Singer and Prashant S. Rao, "U.K. vs. U.S.: How Much of Your Personal Data Can You Get?," *New York Times*, May 20, 2018, https://www.nytimes.com/interactive/2018/05/20/technology/what-data-companies-have-on-you.html.

27. "What Is Responsible Data?," *Responsible Data* (blog), accessed July 28, 2019, https://responsibledata.io/what-is-responsible-data/.

28. This group includes the Center for International Earth Science Information Network (CIESIN), The Earth Institute, Columbia University, City University of New York (CUNY) Institute for Demographic Research (CIDR), The Connectivity Lab at Facebook / Internet.org, ESRI, German Aerospace Center (DLR), European Commission—Joint Research Centre (JRC), ImageCat, Inc., Oak Ridge National Laboratory (ORNL), U.S. Census Bureau, WorldPop, Bill & Melinda Gates Foundation, Google Earth Engine, United Nations Committee of Experts on Global Geospatial Information Management, United Nations Population Division, United Nations Population Fund (UNFPA), and The World Bank.

29. Tobias Tiecke, "Open Population Datasets and Open Challenges—Facebook Code," n.d., https://code.fb.com/core-data/open-population-datasets-and-open-challenges/.

30. Ibid.

31. Shota Iino et al., "Generating High-Accuracy Urban Distribution Map for Short-Term Change Monitoring Based on Convolutional Neural Network by Utilizing SAR Imagery," in *Earth Resources and Environmental Remote Sensing/GIS Applications VIII*, vol. 10428, 2017, 1042803.

32. Interview with Joe Schumacher of CIESIN, March 2018.

33. Ibid.

34. Ibid.

35. Interview with Holly Krambeck, Transportation Specialist World Bank, phone call, August 2018.

36. World Bank, "Open Traffic Data to Revolutionize Transport," December 19, 2019, http://www.worldbank.org/en/news/feature/2016/12/19/open-traffic-data-to-revolutionize-transport.

37. Ibid.

38. World Bank, "World Bank Group and GSMA Announce Partnership to Leverage IoT Big Data for Development," February 26, 2018, http://www.worldbank.org/en/news/press-release/2018/02/26/world-bank-group-and-gsma-announce-partnership-to-leverage-iot-big-data-for-development.

39. Ibid.

40. Robert Kirkpatrick, "A New Type of Philanthropy: Donating Data," *Harvard Business Review*, March 21, 2013, https://hbr.org/2013/03/a-new-type-of-philanthropy-don.

41. Ibid.

42. Vincent D. Blondel et al., "Data for Development: The D4D Challenge on Mobile Phone Data," ArXiv:1210.0137 [cs.CY], September 29, 2012, http://arxiv.org/abs/1210.0137.

43. Amy Wesolowski et al., "Commentary: Containing the Ebola Outbreak—the Potential and Challenge of Mobile Network Data," *PLoS Currents* 6 (September 29, 2014), https://doi.org/10.1371/currents.outbreaks.0177e7fcf52217b8b634376e2f3efc5e; Amy Wesolowski et al., "Quantifying the Impact of Human Mobility on Malaria," *Science* 338, no. 6104 (October 12, 2012): 267–270, https://doi.org/10.1126/science.1223467.

44. Matt Stempeck, "Sharing Data Is a Form of Corporate Philanthropy," *Harvard Business Review*, July 24, 2014, https://hbr.org/2014/07/sharing-data-is-a-form-of-corporate-philanthropy.

45. Mariarosaria Taddeo, "Data Philanthropy and the Design of the Infraethics for Information Societies," *Philosophical Transactions of the Royal Society A* 374, no. 2083 (December 28, 2016): 20160113, https://doi.org/10.1098/rsta.2016.0113.

46. Emmanuel Letouzé, Patrick Vinck, and L Kammourieh, "The Law, Politics and Ethics of Cell Phone Data Analytics," *Data-Pop Alliance White Paper Series. Data-Pop Alliance, World Bank Group, Harvard Humanitarian Initiative, MIT Media Lab and Overseas Development Institute*, April 2015.

BIBLIOGRAPHY

"1842 Sanitary Report." UK Parliament, December 16, 2018. https://www.parliament.uk/about/living-heritage/transformingsociety/livinglearning/coll-9-health1/health-02/1842-sanitary-report-leeds/.

"2014 West Africa Ebola Response—OpenStreetMap Wiki." Accessed January 25, 2019. https://wiki.openstreetmap.org/wiki/2014_West_Africa_Ebola_Response.

Aiello Luca Maria, Schifanella Rossano, Quercia Daniele, and Aletta Francesco. "Chatty Maps: Constructing Sound Maps of Urban Areas from Social Media Data." *Royal Society Open Science* 3, no. 3 (March 1, 2016): 150690. https://doi.org/10.1098/rsos.150690.

Alcock, Susan E., Terence N. D'Altroy, Kathleen D. Morrison, and Carla M. Sinopoli. *Empires: Perspectives from Archaeology and History*. Cambridge: Cambridge University Press, 2001.

Alcorn, Janis B. et al. "Keys to Unleash Mapping's Good Magic." *PLA Notes* 39, no. 2 (2000): 10–13.

Alessa, Ali, and Miad Faezipour. "A Review of Influenza Detection and Prediction through Social Networking Sites." *Theoretical Biology & Medical Modelling* 15 (February 1, 2018). https://doi.org/10.1186/s12976-017-0074-5.

Alexander, Michelle. *The New Jim Crow: Mass Incarceration in the Age of Colorblindness*. New York: The New Press, 2012.

"All the Stops." *Bklynr*. Accessed January 10, 2019. https://www.bklynr.com/all-the-stops/.

Altay, Nezih, and Melissa Labonte. "Challenges in Humanitarian Information Management and Exchange: Evidence from Haiti." *Disasters* 38, no. s1 (2014): S50–72. https://doi.org/10.1111/disa.12052.

Anderson, Margo J. *The American Census: A Social History, Second Edition*. 2nd edition. New Haven: Yale University Press, 2015.

Anselin, Luc, and Sarah Williams. "Digital Neighborhoods." *Journal of Urbanism: International Research on Placemaking and Urban Sustainability* 9, no. 4 (October 1, 2016): 305–328. https://doi.org/10.1080/17549175.2015.1080752.

Arnstein, Sherry. "A Ladder of Citizen Participation." *Journal of the American Institute of Planners* 35, no. 4 (1969): 216–224.

Badger, Emily. "A Census Question That Could Change How Power Is Divided in America." *New York Times*, August 2, 2018. https://www.nytimes.com/2018/07/31/upshot/Census-question-citizenship-power.html.

Bailey, Keiron, and Ted Grossardt. "Toward Structured Public Involvement: Justice, Geography and Collaborative Geospatial/Geovisual Decision Support Systems." *Annals of the Association of American Geographers* 100, no. 1 (2010): 57–86. https://doi.org/10.1080/00045600903364259.

Balazs, Etienne. *Chinese Civilization and Bureaucracy: Variations on a Theme.* New Haven: Yale University Press, 1967.

Baldwin, Davarian L. "Chicago's 'Concentric Circles': Thinking through the Material History of an Iconic Map." In *Many Voices, One Nation: Material Culture Reflections on Race and Migration in the United States,* edited by Margaret Salazar-Porzio, Joan Fragaszy Troyano, and Lauren Safranek, 179–191. Washington, DC: Smithsonian Institution, 2017.

Banbura, Marta, Domenico Giannone, Michele Modugno, and Lucrezia Reichlin. "Now-Casting and the Real-Time Data Flow." *Handbook of Economic Forecasting* 2, no. Part A (2013): 195–237.

Barbaro, Michael, and Tom Zeller Jr. "A Face Is Exposed for AOL Searcher No. 4417749." *New York Times*, August 9, 2006. https://www.nytimes.com/2006/08/09/technology/09aol.html.

Barnes, Trevor J. "Geography's Underworld: The Military–Industrial Complex, Mathematical Modelling and the Quantitative Revolution." *Geoforum* 39, no. 1 (January 2008): 3–16. https://doi.org/10.1016/j.geoforum.2007.09.006.

Bartholomew, Harland. *The Zone Plan.* St. Louis, MO: Nixon-Jones Printing, 1919. https://catalog.hathitrust.org/Record/000342898.

Bartholomew, Harland. *Urban Land Uses.* Cambridge, MA: Harvard University Press, 1932.

Bartholomew, Harland, and City Plan Commission. *Zoning for St. Louis: A Fundamental Part of the City Plan.* St. Louis, MO, 1918. https://ia902708.us.archive.org/5/items/ZoningForSTL/Zoning%20for%20STL.pdf.

Bates, Jo. "The Strategic Importance of Information Policy for the Contemporary Neoliberal State: The Case of Open Government Data in the United Kingdom." *Government Information Quarterly* 31 (2014): 388–395.

Batty, Michael. "At the Crossroads of Urban Growth." *Environment and Planning B: Planning and Design* 41 (2014): 951–953.

Batty, Michael, and Paul M. Torrens. "Modelling and Prediction in a Complex World." Special issue: *Complexity and the Limits of Knowledge* in *Futures* 37, no. 7 (September 1, 2005): 745–66. https://doi.org/10.1016/j.futures.2004.11.003.

Baumgaertner, Emily. "Despite Concerns, Census Will Ask Respondents If They Are U.S. Citizens." *New York Times*, March 28, 2018. https://www.nytimes.com/2018/03/26/us/politics/census-citizenship-question-trump.html.

Bekkers, Victor, and Rebecca Moody. "Visual Events and Electronic Government: What Do Pictures Mean in Digital Government for Citizen Relations?" *Government Information Quarterly* 28, no. 4 (October 1, 2011): 457–465. https://doi.org/10.1016/j.giq.2010.10.006.

Bengtsson, Linus, Xin Lu, Anna Thorson, Richard Garfield, and Johan von Schreeb. "Improved Response to Disasters and Outbreaks by Tracking Population Movements with Mobile Phone Network Data: A Post-Earthquake Geospatial Study in Haiti." *PLoS Medicine* 8, no. 8 (August 30, 2011). https://doi.org/10.1371/journal.pmed.1001083.

Beynon-Davies, Paul. "Significant Threads: The Nature of Data." *International Journal of Information Management* 29, no. 3 (June 1, 2009): 170–188. https://doi.org/10.1016/j.ijinfomgt.2008.12.003.

Bhat, Chandra, Jay Carini, and Rajul Misra. "Modeling the Generation and Organization of Household Activity Stops." *Transportation Research Record: Journal of the Transportation Research Board* 1676 (January 1, 1999): 153–161. https://doi.org/10.3141/1676-19.

Bliss, Laura. "How to Map the Flu." CityLab. Accessed January 26, 2019. https://www.citylab.com/design/2018/01/the-imperfect-science-of-mapping-the-flu/551387/.

Blondel, Vincent D., Markus Esch, Connie Chan, Fabrice Clerot, Pierre Deville, Etienne Huens, Frédéric Morlot, Zbigniew Smoreda, and Cezary Ziemlicki. "Data for Development: The D4D Challenge on Mobile Phone Data." ArXiv:1210.0137, September 29, 2012. http://arxiv.org/abs/1210.0137.

Blum, Sam. "New Data Visualizations Reveal the Racial Disparity of Stop and Frisk." *Brooklyn Magazine*, June 23, 2015. http://www.bkmag.com/2015/06/23/new-data-visualizations-reveal-the-racial-disparity-of-stop-and-frisk/.

Booth, Charles, Richard M. Elman, and Albert Fried. *Charles Booth's London*. New York: Pantheon Books, 1968.

Bostock, Mike, and Ford Fessenden. "'Stop-and-Frisk' Is All but Gone From New York." *New York Times*, September 19, 2014. https://www.nytimes.com/interactive/2014/09/19/nyregion/stop-and-frisk-is-all-but-gone-from-new-york.html.

Bourgault, Jeanne. "How the Global Open Data Movement Is Transforming Journalism." *Wired*, 2013. https://www.wired.com/insights/2013/05/how-the-global-open-data-movement-is-transforming-journalism/.

Bowman, John S. *Columbia Chronologies of Asian History and Culture*. New York: Columbia University Press, 2000.

boyd, danah, and Kate Crawford. "Critical Questions for Big Data: Provocations for a Cultural, Technological, and Scholarly Phenomenon." *Information, Communication & Society* 15, no. 5 (2012): 662–679.

Boyer, Christine M. *Dreaming the Rational City: The Myth of American City Planning*. Cambridge, MA: MIT Press, 1986.

Brown, Azby, Pieter Franken, Sean Bonner, Nick Dolezal, and Joe Moross. "Safecast: Successful Citizen-Science for Radiation Measurement and Communication after Fukushima." *Journal of Radiological Protection* 36, no. 2 (2016): S82.

Brownstein, John S., Clark C. Freifeld, Ben Y. Reis, and Kenneth D. Mandl. "Surveillance Sans Frontières: Internet-Based Emerging Infectious Disease Intelligence and the HealthMap Project." *PLOS Medicine* 5, no. 7 (July 8, 2008): e151. https://doi.org/10.1371/journal.pmed.0050151.

Bulmer, Martin. *The Chicago School of Sociology: Institutionalization, Diversity, and the Rise of Sociological Research*. Chicago: University of Chicago Press, 1986.

Bulmer, Martin, Kevin Bales, and Kathryn Kish Sklar. *The Social Survey in Historical Perspective, 1880–1940*. Cambridge: Cambridge University Press, 1991.

Bulmer, Martin, Kevin Bales, Kathryn Kish Sklar, and Kathryn Kish Sklar. *The Social Survey in Historical Perspective, 1880–1940*. Cambridge: Cambridge University Press, 1991.

Burgess, Ernest Watson. "Map of the Radial Expansion and the Five Urban Zones," n.d. Ernest Watson Burgess Papers. University of Chicago Library.

Burke, Jeffrey A., Deborah Estrin, Mark Hansen, Andrew Parker, Nithya Ramanathan, Sasank Reddy, and Mani B. Srivastava. "Participatory Sensing," UCLA Center for Embedded Network Sensing, 2006, https://escholarship.org/uc/item/19h777qd.

Burns, Ryan. "Moments of Closure in the Knowledge Politics of Digital Humanitarianism." *Geoforum* 53 (May 1, 2014): 51–62. https://doi.org/10.1016/j.geoforum.2014.02.002.

Busari, Stephanie. "Tweeting the Terror: How Social Media Reacted to Mumbai." CNN.com, 2008. https://www.cnn.com/2008/WORLD/asiapcf/11/27/mumbai.twitter/.

Byington, Margaret Frances, Paul Underwood Kellogg, and Russell Sage Foundation. *Homestead: The Households of a Mill Town*, 1910. https://archive.org/details/homesteadhouseho00byinuoft/page/n189.

"Case Study: *NYC v Floyd*." Catalysts for Collaboration. Accessed January 10, 2019. https://catalystsforcollaboration.org/casestudy/nycfloyd.html.

Castells, Manuel. *Communication Power*. Oxford: Oxford University Press, 2013.

Chambers, Robert. "Participatory Mapping and Geographic Information Systems: Whose Map? Who Is Empowered and Who Disempowered? Who Gains and Who Loses?" *Electronic Journal of Information Systems in Developing Countries* 25, no. 1 (2006): 1–11.

Chambers, Robert. "The Origins and Practice of Participatory Rural Appraisal." *World Development* 22, no. 7 (July 1, 1994): 953–969. https://doi.org/10.1016/0305-750X(94)90141-4.

Chi, Guanghua, Yu Liu, Zhengwei Wu, and Haishan Wu. "Ghost Cities Analysis Based on Positioning Data in China." ArXiv:1510.08505, October 28, 2015. http://arxiv.org/abs/1510.08505.

Chunara, Rumi, Jason R. Andrews, and John S. Brownstein. "Social and News Media Enable Estimation of Epidemiological Patterns Early in the 2010 Haitian Cholera Outbreak." *American Journal of Tropical Medicine and Hygiene* 86, no. 1 (January 1, 2012): 39–45. https://doi.org/10.4269/ajtmh.2012.11-0597.

City of New York Board of Estimate and Apportionment. *Use District Map*. New York: City of New York, 1916. https://digitalcollections.nypl.org/items/510d47e4-80df-a3d9-e040-e00a18064a99.

City Plan Commission. *Map of the City of St. Louis: Distribution of Negro Population, Census of 1930. (Realtors' Red Line Map.)*. 1934. Map. http://collections.mohistory.org/resource/221591.

Clougherty Jane E., and Kubzansky Laura D. "A Framework for Examining Social Stress and Susceptibility to Air Pollution in Respiratory Health." *Environmental Health Perspectives* 117, no. 9 (September 1, 2009): 1351–1358. https://doi.org/10.1289/ehp.0900612.

Cook, Samantha, Corrie Conrad, Ashley L. Fowlkes, and Matthew H. Mohebbi. "Assessing Google Flu Trends Performance in the United States during the 2009 Influenza Virus A (H1N1) Pandemic." *PLOS ONE* 6, no. 8 (August 19, 2011): e23610. https://doi.org/10.1371/journal.pone.0023610.

"Correcting the *New York Times* on College 'Enrollment Gaps.'" AEI, June 25, 2013. http://www.aei.org/publication/correcting-the-new-york-times-on-college-enrollment-gaps/.

Couldry, Nick. *The Costs of Connection: How Data Is Colonizing Human Life and Appropriating It for Capitalism*. 1st edition. Stanford, CA: Stanford University Press, 2019.

Crampton, Jeremy W. "Maps as Social Constructions: Power, Communication and Visualization." *Progress in Human Geography* 25, no. 2 (2001): 235–252.

Crampton, Jeremy W., and John Krygier. "An Introduction to Critical Cartography." *ACME: An International E-Journal for Critical Geographies* 4, no. 1 (2006): 11–33.

Cranshaw, Justin, Raz Schwartz, Jason Hong, and Norman Sadeh. "The Livehoods Project: Utilizing Social Media to Understand the Dynamics of a City." *Proceedings of the Sixth International AAAI Conference on Weblogs and Social Media*, 2012.

Cravero, Paolo. "Mapping for Food Safety." International Institute for Environment and Development, December 21, 2015. https://www.iied.org/mapping-for-food-safety.

Crawford, Kate, and Megan Finn. "The Limits of Crisis Data: Analytical and Ethical Challenges of Using Social and Mobile Data to Understand Disasters." *GeoJournal* 80, no. 4 (August 1, 2015): 491–502. https://doi.org/10.1007/s10708-014-9597-z.

Crean, Sarah. "While Improving, City's Air Quality Crisis Quietly Persists." Gotham Gazette, June 19, 2014. http://www.gothamgazette.com/government/5111-while-improving-quiet-crisis-air-quality-persists-new-york-city-asthma-air-pollution.

Crowd & The Cloud, The. "DIY Science for the People: Q&A with Public Lab's Jeff Warren." *Medium* (blog), March 27, 2017. https://medium.com/@crowdandcloud/diy-science-for-the-people-q-a-with-public-labs-jeff-warren-c45b6689a228.

Currid, Elizabeth, and Sarah Williams. "The Geography of Buzz: Art, Culture and the Social Milieu in Los Angeles and New York." *Journal of Economic Geography* 10, no. 3 (2009): 423–451.

Daley, David. "The Secret Files of the Master of Modern Republican Gerrymandering." *New Yorker*, September 6, 2019. https://www.newyorker.com/news/news-desk/the-secret-files-of-the-master-of-modern-republican-gerrymandering

Data Revolution Group. "A World That Counts." Data Revolution Group, 2014. http://www.undatarevolution.org/wp-content/uploads/2014/11/A-World-That-Counts.pdf.

DataMade. "Chicago's Million Dollar Blocks." Chicago's Million Dollar Blocks, January 7, 2019. https://chicagosmilliondollarblocks.com/.

Deegan, Mary Jo. "W. E. B. Du Bois and the Women of Hull-House, 1895–1899." *American Sociologist* 19, no. 4 (December 1, 1988): 301–311. https://doi.org/10.1007/BF02691827.

Denvir, Daniel. "The Key Ingredient in Stop and Frisk Reform: Open Data." CityLab, August 25, 2015. http://www.citylab.com/crime/2015/08/the-missing-ingredient-in-stop-and-frisk-accountability-open-data/402026/.

D'Ignazio, Catherine, and Lauren F. Klein. *Data Feminism*. Cambridge, MA: MIT Press, 2020.

"Diagram of a Slave Ship," 1801. https://www.bl.uk/learning/timeline/item106661.html.

Ding, W., S. Zheng, and X. Guo. "Value of Access to Jobs and Amenities: Evidence from New Residential Properties in Beijing." *Tsinghua Science and Technology* 15, no. 5 (October 2010): 595–603. https://doi.org/10.1016/S1007-0214(10)70106-1.

"Do Affirmative Action Bans Hurt Minority Students?" IVN.us, April 29, 2014. https://ivn.us/2014/04/29/affirmative-action-bans-hurt-minority-students-states-show-mixed-results/.

Du Bois, W. E. B. *[A Series of Statistical Charts Illustrating the Condition of the Descendants of Former African Slaves Now in Residence in the United States of America] Negro Business Men in the United States.* Circa 1900. 1 drawing: ink and watercolor, 710 x 560 mm. LOT 11931, no. 57 (M) [P&P]. Library of Congress Prints and Photographs Division. https://www.loc.gov/pictures/item/2014645363/.

Du Bois, W. E. B. *[The Georgia Negro] Income and Expenditure of 150 Negro Families in Atlanta, Ga., U.S.A.* Circa 1900. Drawing: ink, watercolor, and photographic print, 710 x 560 mm (board). LOT 11931, no. 31 [P&P]. Library of Congress Prints and Photographs Division. https://www.loc.gov/pictures/item/2013650354/.

Du Bois, William Edward Burghardt, and Isabel Eaton. *The Philadelphia Negro: A Social Study.* 14. Published for the University, 1899.

Duggan, Maeve, and Joanna Brenner. *The Demographics of Social Media Users, 2012.* Vol. 14. Pew Research Center's Internet & American Life Project Washington, DC, 2013.

Duhigg, Charles. "How Companies Learn Your Secrets." *New York Times,* February 16, 2012. https://www.nytimes.com/2012/02/19/magazine/shopping-habits.html.

Earle, Paul S., Daniel C. Bowden, and Michelle Guy. "Twitter Earthquake Detection: Earthquake Monitoring in a Social World." *Annals of Geophysics* 54, no. 6 (January 14, 2012). https://doi.org/10.4401/ag-5364.

Eastman, Crystal. *Work-Accidents and the Law,* 1910. https://archive.org/details/cu31924019223035/page/n231.

Eglash, Ron, Juan E. Gilbert, and Ellen Foster. "Toward Culturally Responsive Computing Education." *Communications of the ACM* 56, no. 7 (July 2013): 33–36. https://doi.org/10.1145/2483852.2483864.

Elmer, Greg, Ganaele Langlois, and Joanna Redden. *Compromised Data: From Social Media to Big Data.* New York: Bloomsbury Publishing, 2015.

Elwood, Sarah. "Critical Issues in Participatory GIS: Deconstructions, Reconstructions, and New Research Directions." *Transactions in GIS* 10, no. 5 (2006): 693–708.

Elwood, Sarah A. "GIS Use in Community Planning: A Multidimensional Analysis of Empowerment." *Environment and Planning A* 34, no. 5 (May 2002): 905–22. https://doi.org/10.1068/a34117.

Elwood, Sarah, and Helga Leitner. "GIS and Spatial Knowledge Production for Neighborhood Revitalization: Negotiating State Priorities and Neighborhood Visions." *Journal of Urban Affairs* 25, no. 2 (2003): 139–157. https://doi.org/10.1111/1467-9906.t01-1-00003.

Estrin, Deborah, K. Mani Chandy, R. Michael Young, Larry Smarr, Andrew Odlyzko, David Clark, Viviane Reding, Toru Ishida, Sharad Sharma, and Vinton G. Cerf. "Participatory Sensing: Applications and Architecture [Internet Predictions]." *IEEE Internet Computing* 14, no. 1 (2010): 12–42.

European Union. "2018 Reform of EU Data Protection Rules," 2018. https://ec.europa.eu/commission/priorities/justice-and-fundamental-rights/data-protection/2018-reform-eu-data-protection-rules_en.

Evans, Gary W. *Environmental Stress.* Cambridge: Cambridge University Press, 1984.

Fagan, Jeffrey, and Amanda Geller. "Following the Script: Narratives of Suspicion in Terry Stops in Street Policing." SSRN Scholarly Paper. Rochester, NY: Social Science Research Network, August 1, 2014. https://papers.ssrn.com/abstract=2485375.

Fainstein, Susan, and DeFilippis, James. *Readings in Planning Theory*. Hoboken, NJ: John Wiley & Sons, 2015.

Faludi, A. *A Reader in Planning Theory*. New York: Pergamon Press, 1973.

Fan, Maureen. "Gray Wall Dims Hopes of 'Green' Games; China Has Vowed to Curb Pollution Before '08 Olympics, but Its Secrecy Is Feeding Skepticism." *Washington Post*, October 16, 2007.

Fan, Maureen. "If That Doesn't Clear the Air . . . China, Struggling to Control Smog, Announces 'Just-in-Case' Plan." *Washington Post*, July 31, 2008.

"Federal Court Rules Three Texas Congressional Districts Illegally Drawn." NPR.org. Accessed January 30, 2019. https://www.npr.org/sections/thetwo-way/2017/03/11/519839892/federal-court-rules-three-texas-congressional-districts-illegally-drawn.

Felstiner, Alek. "Working the Crowd: Employment and Labor Law in the Crowdsourcing Industry." *Berkeley Journal of Employment and Labor Law* 32 (2011): 143–204.

Fessenden, Ford, and Josh Keller. "How Minorities Have Fared in States With Affirmative Action Bans," January 7, 2019. https://www.nytimes.com/interactive/2013/06/24/us/affirmative-action-bans.html.

Forester, John. *Planning in the Face of Power*. Oakland: University of California Press, 1988.

Forester, John. *The Deliberative Practitioner: Encouraging Participatory Planning Processes*. Cambridge, MA: MIT Press, 1999.

Form-Based Codes Institute. "Zoning for Equity: Raising All Boats—Form-Based Codes Institute at Smart Growth America: Form-Based Codes Institute at Smart Growth America," 2019. https://formbasedcodes.org/blog/zoning-equity-raising-boats/.

"Freeway History: Overtown." MIAMI, January 3, 2019. http://miamiplanning.weebly.com/freeway-history--overtown.html.

Freire, Paulo. *Pedagogy of the Oppressed*. New York: Bloomsbury Publishing, 2018.

Gabrys, Jennifer. *Program Earth: Environmental Sensing Technology and the Making of a Computational Planet*. Vol. 49. Minneapolis: University of Minnesota Press, 2016.

Gal-Tzur, Ayelet, Susan M. Grant-Muller, Tsvi Kuflik, Einat Minkov, Silvio Nocera, and Itay Shoor. "The Potential of Social Media in Delivering Transport Policy Goals." *Transport Policy* 32 (2014): 115–123. https://doi.org/10.1016/j.tranpol.2014.01.007.

Gauff Consultants. *Nairobi / Proposed MRTS Commuter Rail Network*. 2014.

Geddes, Patrick. *Cities in Evolution*. London: William and Norgate Limited, 1949.

Geddes, Sir Patrick. *Town Planning in Kapurthala: A Report to the HH the Maharaja of Kapurthala*, 1917.

Ghose, Rina. "Politics of Scale and Networks of Association in Public Participation GIS." *Environment and Planning A* 39, no. 8 (2007): 1961–1980.

Giannone, Domenico, Lucrezia Reichlin, and David Small. "Nowcasting: The Real-Time Informational Content of Macroeconomic Data." *Journal of Monetary Economics* 55, no. 4 (May 2008): 665–676. https://doi.org/10.1016/j.jmoneco.2008.05.010.

Gidley, Ben. *The Proletarian Other: Charles Booth and the Politics of Representation*. London: Goldsmiths University of London, 2000.

Ginsberg, Jeremy, Matthew H. Mohebbi, Rajan S. Patel, Lynnette Brammer, Mark S. Smolinski, and Larry Brilliant. "Detecting Influenza Epidemics Using Search Engine Query Data." *Nature* 457, no. 7232 (February 2009): 1012–1014. https://doi.org/10.1038/nature07634.

Gitelman, Lisa. *"Raw Data" Is an Oxymoron*. Cambridge, MA: MIT Press, 2013.

Glaeser, Edward, Wei Huang, Yueran Ma, and Andrei Shleifer. "A Real Estate Boom with Chinese Characteristics." *Journal of Economic Perspectives* 31, no. 1 (2017): 93–116.

Glaeser, Edward L., Jed Kolko, and Albert Saiz. "Consumer City." *Journal of Economic Geography* 1, no. 1 (2001): 27–50. https://doi.org/10.1093/jeg/1.1.27.

Glass, James J. "Citizen Participation in Planning: The Relationship Between Objectives and Techniques." *Journal of the American Planning Association* 45, no. 2 (April 1979): 180–189. https://doi.org/10.1080/01944367908976956.

Goldsmith, Stephen, and Crawford, Susan. *The Responsive City: Engaging Communities Through Data-Smart Governance*. Hoboken, NJ: John Wiley & Sons, 2014.

Goldstein, Brett, and Lauren Dyson, eds. *Beyond Transparency: Open Data and the Future of Civic Innovation*. San Francisco: Code for America Press, 2013.

Goodchild, Michael F. "Citizens as Sensors: The World of Volunteered Geography." *GeoJournal* 69, no. 4 (2007): 211–221.

Google. "Google Maps Tiles," n.d. via Safecast.org.

Gordon, Eric, Steven Schirra, and Justin Hollander. "Immersive Planning: A Conceptual Model for Designing Public Participation with New Technologies." *Environment and Planning B: Planning and Design* 38, no. 3 (June 2011): 505–519. https://doi.org/10.1068/b37013.

Gratz, Roberta Brandes. *The Battle for Gotham: New York in the Shadow of Robert Moses and Jane Jacobs*. New York: Nation Books, 2010.

Gray, James N., David Pessel, and Pravin P. Varaiya. "A Critique of Forrester's Model of an Urban Area." *IEEE Transactions on Systems, Man, and Cybernetics*, no. 2 (1972): 139–144.

Greenfield, Patrick. "The Cambridge Analytica Files: The Story So Far." *Guardian*, March 25, 2018. https://www.theguardian.com/news/2018/mar/26/the-cambridge-analytica-files-the-story-so-far.

Grein, T. W., K. B. Kamara, G. Rodier, A. J. Plant, P. Bovier, M. J. Ryan, T. Ohyama, and D. L. Heymann. "Rumors of Disease in the Global Village: Outbreak Verification." *Emerging Infectious Diseases* 6, no. 2 (2000): 97–102.

Groner, Anya. "Healing the Gulf with Buckets and Balloons." Guernica, September 8, 2016. https://www.guernicamag.com/anya-groner-healing-the-gulf-with-buckets-and-balloons/.

Guagliardo, Mark F. "Spatial Accessibility of Primary Care: Concepts, Methods and Challenges." *International Journal of Health Geographics* 3, no. 1 (February 26, 2004): 3. https://doi.org/10.1186/1476-072X-3-3.

Guagliardo, Mark F, Cynthia R Ronzio, Ivan Cheung, Elizabeth Chacko, and Jill G Joseph. "Physician Accessibility: An Urban Case Study of Pediatric Providers." *Health & Place* 10, no. 3 (September 1, 2004): 273–283. https://doi.org/10.1016/j.healthplace.2003.01.001.

Hafner, Katie. "Researchers Yearn to Use AOL Logs, but They Hesitate." *New York Times*, August 23, 2006. https://www.nytimes.com/2006/08/23/technology/23search.html.

Hall, Peter. "The Turbulent Eighth Decade: Challenges to American City Planning." *Journal of the American Planning Association* 55, no. 3 (September 30, 1989): 275–282. https://doi.org/10.1080/01944368908975415.

Hammack, David C., and Stanton Wheeler. *Social Science in the Making: Essays on the Russell Sage Foundation, 1907–1972*. Russell Sage Foundation, 1995.

Handy, Susan L. "Regional Versus Local Accessibility: Neo-Traditional Development and Its Implications for Non-Work Travel on JSTOR." *Built Environment* 18, no. 4 (1992): 253–267.

Hansen, Walter G. "How Accessibility Shapes Land Use." *Journal of the American Institute of Planners* 25, no. 2 (May 1, 1959): 73–76. https://doi.org/10.1080/01944365908978307.

Hao, Jiming, and Litao Wang. "Improving Urban Air Quality in China: Beijing Case Study." *Journal of the Air & Waste Management Association* 55, no. 9 (2005): 1298–1305.

Hardy, Anne. "'Death Is the Cure of All Diseases': Using the General Register Office Cause of Death Statistics for 1837–1920." *Social History of Medicine* 7, no. 3 (December 1, 1994): 472–492. https://doi.org/10.1093/shm/7.3.472.

Harley, J. Brian. "Maps, Knowledge, and Power." *Geographic Thought: A Praxis Perspective*, 2009, 129–148.

Harley, J. Brian. "Maps, Knowledge, and Power." *Geographic Thought: A Praxis Perspective*, 2009, 129–148.

Harris, Trevor, and Daniel Weiner. "Empowerment, Marginalization, and 'Community-Integrated' GIS." *Cartography and Geographic Information Systems* 25, no. 2 (1998): 67–76. https://doi.org/10.1559/152304098782594580.

Harrison, A. *Economics and Land Use Planning*. London: Croom Helm, 1977.

Harrison, Roy M., and Jianxin Yin. "Particulate Matter in the Atmosphere: Which Particle Properties Are Important for Its Effects on Health?" *Science of the Total Environment* 249, no. 1 (April 17, 2000): 85–101. https://doi.org/10.1016/S0048-9697(99)00513-6.

Harvey, David. "On the History and Present Condition of Geography: An Historical Materialist Manifesto." *Professional Geographer* 36, no. 1 (February 1, 1984): 1–11. https://doi.org/10.1111/j.0033-0124.1984.00001.x.

Harvey, Francis. "To Volunteer or to Contribute Locational Information? Towards Truth in Labeling for Crowdsourced Geographic Information." In *Crowdsourcing Geographic Knowledge: Volunteered Geographic Information (VGI) in Theory and Practice*, edited by Daniel Sui, Sarah Elwood, and Michael Goodchild, 31–42. Dordrecht: Springer Netherlands, 2013. https://doi.org/10.1007/978-94-007-4587-2_3.

Haselton, Todd. "Google Still Keeps a List of Everything You Ever Bought Using Gmail, Even If You Delete All Your Emails." CNBC, July 5, 2019. https://www.cnbc.com/2019/07/05/google-gmail-purchase-history-cant-be-deleted.html.

Hecht, Brent, and Monica Stephens. "A Tale of Cities: Urban Biases in Volunteered Geographic Information," *ICWSM* 14 (2014): 197–205.

Helft, Miguel. "Google Uses Web Searches to Track Flu's Spread." *New York Times*, November 11, 2008. https://www.nytimes.com/2008/11/12/technology/internet/12flu.html.

Hess, Daniel Baldwin. "Access to Employment for Adults in Poverty in the Buffalo-Niagara Region." *Urban Studies* 42, no. 7 (June 1, 2005): 1177–1200. https://doi.org/10.1080/00420980500121384.

Heymann, David L, and Guénaël R Rodier. "Hot Spots in a Wired World: WHO Surveillance of Emerging and Re-Emerging Infectious Diseases." *Lancet Infectious Diseases* 1, no. 5 (December 1, 2001): 345–353. https://doi.org/10.1016/S1473-3099(01)00148-7.

Hillier, Amy E. "Redlining and the Home Owners' Loan Corporation." *Journal of Urban History* 29, no. 4 (May 1, 2003): 394–420. https://doi.org/10.1177/0096144203029004002.

Hodgson, Dorothy L., and Richard A. Schroeder. "Dilemmas of Counter-Mapping Community Resources in Tanzania." *Development and Change* 33, no. 1 (2002): 79–100. https://doi.org/10.1111/1467-7660.00241.

Horner, Mark W. "Spatial Dimensions of Urban Commuting: A Review of Major Issues and Their Implications for Future Geographic Research." *Professional Geographer* 56, no. 2 (May 1, 2004): 160–173. https://doi.org/10.1111/j.0033-0124.2004.05602002.x.

Howe, Jeff. "The Rise of Crowdsourcing." *Wired*, June 1, 2006. https://www.wired.com/2006/06/crowds/.

Huang, Daisy J., Charles K. Leung, and Baozhi Qu. "Do Bank Loans and Local Amenities Explain Chinese Urban House Prices?" *China Economic Review* 34 (July 1, 2015): 19–38. https://doi.org/10.1016/j.chieco.2015.03.002.

Huhn, Wilson. *Schuette v. Coalition to Defend Affirmative Action*. US Supreme Court, No. 572 U.S. 2013.

"Humanitarian Tracker." Humanitarian Tracker, October 15, 2018. https://www.humanitariantracker.org.

"IBM Smarter Cities—Future Cities—United States," August 3, 2016. http://www.ibm.com/smarterplanet/us/en/smarter_cities/overview/.

"Identity Resolution Service & Data Onboarding." LiveRamp. Accessed July 13, 2019. https://liveramp.com/.

IDH Shepherd. "Mapping the Poor in Late-Victorian London: A Multi-Scale Approach." In *Getting the Measure of Poverty*, 148–176. London: Aldershot, Ashgate, 1999.

IEAG, UN. "A World That Counts–Mobilising the Data Revolution for Sustainable Development." New York: United Nations, 2014.

Iino, Shota, Riho Ito, Kento Doi, Tomoyuki Imaizumi, and Shuhei Hikosaka. "Generating High-Accuracy Urban Distribution Map for Short-Term Change Monitoring Based on Convolutional Neural Network by Utilizing SAR Imagery." In *Earth Resources and Environmental Remote Sensing/GIS Applications VIII*, 10428:1042803, 2017.

Incite! and Women of Color against Violence. *The Revolution Will Not Be Funded: Beyond the Non-Profit Industrial Complex*. South End Press, 2007.

Inka Khipu (Fiber Recording Device). Circa 1400–1532. Cords, knotted and twisted; cotton and wool, Overall: 85 x 108 cm (33 7/16 x 42 1/2 in.). https://clevelandart.org/art/1940.469.

Interview with Iyad Kheirbek of [New York City Department of Mental Health and Hygiene], January 2015.

Jacobs, Jane. *The Death and Life of Great American Cities*. New York: Random House, 1961.

"James Madison to W. T. Barry, August 4, 1822." Image. Library of Congress, Washington, DC. Accessed January 26, 2019. https://www.loc.gov/resource/mjm.20_0155_0159/?st=list.

Jin, Xiaobin, Ying Long, Wei Sun, Yuying Lu, Xuhong Yang, and Jingxian Tang. "Evaluating Cities' Vitality and Identifying Ghost Cities in China with Emerging Geographical Data." *Cities* 63 (March 1, 2017): 98–109. https://doi.org/10.1016/j.cities.2017.01.002.

Jonsson, Patrik. "Gulf Oil Spill: Al Gore Slams BP for Lack of Media Access." *Christian Science Monitor*, June 15, 2010. https://www.csmonitor.com/USA/Politics/2010/0615/Gulf-oil-spill-Al-Gore-slams-BP-for-lack-of-media-access.

Joyce, Patrick. *The Rule of Freedom: Liberalism and the Modern City*. London: Verso, 2003.

Kanter, James. "Google Showcases Augmented-Reality Navigation on Google Maps—Business Insider." *Business Insider*, May 9, 2018. https://www.businessinsider.com/google-showcases-augmented-reality-navigation-on-google-maps-2018-5.

Kass-Hout, Taha A., and Hend Alhinnawi. "Social Media in Public Health." *British Medical Bulletin* 108, no. 1 (2013): 5–24.

Keefe, John. "Stop & Frisk | Guns." *WNYC*, 2013. https://project.wnyc.org/stop-frisk-guns/.

Kellog, Paul Underwood. "Figure 1.1." In *The Pittsburgh Survey: Findings in Six Volumes*, Ed. 1909. 112 East 64th Street, New York, NY 10065: © Russell Sage Foundation, n.d.

Kellogg, Paul Underwood. *The Pittsburgh Survey; Findings in Six Volumes*. Vol. 5, 1909. https://archive.org/details/pittsburghsurvey05kelluoft/page/xvi.

Kellogg, Wendy A. "From the Field: Observations on Using GIS to Develop a Neighborhood Environmental Information System for Community-Based Organizations" 11, no. 1 (1999): 19.

Kimball, Miles A. "London through Rose-Colored Graphics: Visual Rhetoric and Information Graphic Design in Charles Booth's Maps of London Poverty." *Journal of Technical Writing and Communication* 36, no. 4 (October 2006): 353–381. https://doi.org/10.2190/K561-40P2-5422-PTG2.

Kinney, Patrick L., Maneesha Aggarwal, Mary E. Northridge, Nicole A. Janssen, and Peggy Shepard. "Airborne Concentrations of PM (2.5) and Diesel Exhaust Particles on Harlem Sidewalks: A Community-Based Pilot Study." *Environmental Health Perspectives* 108, no. 3 (2000): 213.

Kirkpatrick, Robert. "A New Type of Philanthropy: Donating Data." *Harvard Business Review* (March 21, 2013). https://hbr.org/2013/03/a-new-type-of-philanthropy-don.

Kitchin, Rob. "The Real-Time City? Big Data and Smart Urbanism." *GeoJournal* 79, no. 1 (February 1, 2014): 1–14. https://doi.org/10.1007/s10708-013-9516-8.

"KitHub Kits for Learning about Environmental Monitoring and STEAM." KitHub. Accessed January 25, 2019. https://kithub.cc/.

Klembczyk, Joseph Jeffrey, Mehdi Jalalpour, Scott Levin, Raynard E. Washington, Jesse M. Pines, Richard E. Rothman, and Andrea Freyer Dugas. "Google Flu Trends Spatial Variability Validated Against Emergency Department Influenza-Related Visits." *Journal of Medical Internet Research* 18, no. 6 (2016): e175. https://doi.org/10.2196/jmir.5585.

Koti, Francis, and Daniel Weiner. "(Re) Defining Peri-Urban Residential Space Using Participatory GIS in Kenya." *Electronic Journal of Information Systems in Developing Countries* 25, no. 1 (June 2006): 1–12. https://doi.org/10.1002/j.1681-4835.2006.tb00169.x.

Kovacs-Gyori, Anna, Alina Ristea, Clemens Havas, Bernd Resch, and Pablo Cabrera-Barona. "#London2012: Towards Citizen-Contributed Urban Planning Through Sentiment Analysis of Twitter Data." *Urban Planning* 3, no. 1 (2018): 75–99. https://doi.org/10.17645/up.v3i1.1287.

Krambeck, Holly. Interview with Holly Krambeck, transportation specialist, World Bank. Phone call, August 2018.

Krambeck, Holly. "The Open Transport Partnership." presented at the Transforming Transportation, Washington, DC. January 20, 2017. https://www.slideshare.net/EMBARQNetwork/the-open-transport-partnership.

Kurgan, Laura. *Close Up at a Distance: Mapping, Technology, and Politics*. Cambridge, MA: MIT Press, 2013.

Kwan, Mei-Po. "Gender, the Home-Work Link, and Space-Time Patterns of Nonemployment Activities." *Economic Geography* 75, no. 4 (October 1, 1999): 370–394. https://doi.org/10.1111/j.1944-8287.1999.tb00126.x.

"Landmark Decision: Judge Rules NYPD Stop and Frisk Practices Unconstitutional, Racially Discriminatory." Center for Constitutional Rights. Accessed July 21, 2019. https://ccrjustice.org/node/1269.

Lansley, Guy, and Longley, Paul A. "The Geography of Twitter Topics in London." *Computers, Environment and Urban Systems* 58 (2016): 85–96.

Lartey, Jamiles. "US Police Killings Undercounted by Half, Study Using Guardian Data Finds." *Guardian*, October 11, 2017. https://www.theguardian.com/us-news/2017/oct/11/police-killings-counted-harvard-study.

Latham, Annabel. "Cambridge Analytica Scandal: Legitimate Researchers Using Facebook Data Could Be Collateral Damage." The Conversation. Accessed July 6, 2019. http://theconversation.com/cambridge-analytica-scandal-legitimate-researchers-using-facebook-data-could-be-collateral-damage-93600.

Lazar, Nicole, and Christine Franklin. "The Big Picture: Preparing Students for a Data-Centric World." *Chance* 28, no. 4 (2015): 43–45.

Lazer, David, Ryan Kennedy, Gary King, and Alessandro Vespignani. "The Parable of Google Flu: Traps in Big Data Analysis." *Science* 343, no. 6176 (March 14, 2014): 1203–1205. https://doi.org/10.1126/science.1248506.

Lee, Douglass B. "Requiem for Large-Scale Models." *Journal of the American Institute of Planners* 39, no. 3 (May 1973): 163–178. https://doi.org/10.1080/01944367308977851.

Letouzé, Emmanuel, Patrick Vinck, and L Kammourieh. "The Law, Politics and Ethics of Cell Phone Data Analytics." *Data-Pop Alliance White Paper Series*. Data-Pop Alliance, World Bank Group, Harvard Humanitarian Initiative, MIT Media Lab and Overseas Development Institute. April 2015.

Levinson, David M. "Accessibility and the Journey to Work." *Journal of Transport Geography* 6, no. 1 (March 1, 1998): 11–21. https://doi.org/10.1016/S0966-6923(97)00036-7.

Levitt, Justin. "A Citizen's Guide to Redistricting." SSRN Scholarly Paper. Rochester, NY: Social Science Research Network, June 28, 2008. https://papers.ssrn.com/abstract=1647221.

Levy, Steven. *Hackers: Heroes of the Computer Revolution*. Vol. 14. Garden City, NY: Anchor Press/Doubleday 1984.

Light, Jennifer S. *From Warfare to Welfare: Defense Intellectuals and Urban Problems in Cold War America*. Baltimore: Johns Hopkins University Press, 2003.

Lin, Wen. "COUNTER-CARTOGRAPHIES." *Introducing Human Geographies*, 2013, 215.

"List of Physical Visualizations." Accessed January 27, 2019. http://dataphys.org/list/stop-and-frisk-physical-data-filtering/.

"Livehoods: New York." Accessed August 2, 2019. http://livehoods.org/maps/nyc.

"London and the Industrial Revolution." *Londontopia* (blog). Accessed June 27, 2019. https://londontopia.net/history/london-industrial-revolution/.

London School of Economics and Political Science. "Charles Booth's London." Accessed August 2, 2019. https://booth.lse.ac.uk/map/13/-0.1565/51.5087/100/0.

Lovelace, Eldridge. *Harland Bartholomew: His Contribution to American Urban Planning*. Urbana: Department of Urban and Regional Planning, University of Illinois, 1993.

Lu, Xin, Linus Bengtsson, and Petter Holme. "Predictability of Population Displacement after the 2010 Haiti Earthquake." *Proceedings of the National Academy of Sciences of the United States of America* 109, no. 29 (July 17, 2012): 11576–11581. https://doi.org/10.1073/pnas.1203882109.

Luo, Wei, and Fahui Wang. "Measures of Spatial Accessibility to Health Care in a GIS Environment: Synthesis and a Case Study in the Chicago Region." *Environment and Planning B: Planning and Design* 30, no. 6 (December 1, 2003): 865–884. https://doi.org/10.1068/b29120.

Lynch, Kevin. *Some Major Problems, Boston : MIT Libraries*. 1960. Perceptual Map. MC 208, Box 6. MIT Dome. https://dome.mit.edu/handle/1721.3/36515.

Lynch, Kevin. *The Image of the City*. Cambridge, MA: MIT Press, 1960.

@ma3route. "Ma3Route Is a Mobile/Web/SMS Platform That Crowd-Sources for Transport Data and Provides Users with Information on Traffic, Matatu Directions and Driving Reports." Accessed August 1, 2019. http://www.ma3route.com/.

Maciag, Mike. "How Driverless Cars Could Be a Big Problem for Cities." Governing, 2017. https://www.governing.com/topics/finance/gov-cities-traffic-parking-revenue-driverless-cars.html.

Mann, Gideon. "Private Data and the Public Good," May 17, 2016. https://medium.com/@gideonmann/private-data-and-the-public-good-9c94c656ff28.

MapMill. *Image of Map Mill App Courtesy of PublicLab.Org*. n.d. Website.

"Mapping Vice in San Francisco—Mapping the Nation Blog." Accessed June 29, 2019. http://www.mappingthenation.com/blog/mapping-vice-in-san-francisco/.

"Maps of Gilded Age San Francisco, Chicago, and New York—Mapping the Nation Blog." Accessed June 29, 2019. http://www.mappingthenation.com/blog/maps-of-gilded-age-san-francisco-chicago-and-new-york/.

McDonald, Sean M. "Ebola: A Big Data Disaster-Privacy, Property, and the Law of Disaster Experimentation." *Centre for Internet and Society*, no. 2016.01 (2016).

McNamara, Amelia, and Mark Hansen. "Teaching Data Science to Teenagers." In *Proceedings of the Ninth International Conference on Teaching Statistics*, 2014.

McPhail, Thomas L., and Everett M. Rogers. *Electronic Colonialism: The Future of International Broadcasting and Communication*. Sage Beverly Hills, 1981.

Meares, Tracey. "Programming Errors: Understanding the Constitutionality of Stop and Frisk as a Program, Not an Incident," 2014. https://doi.org/10.13140/2.1.4252.6404.

Meier, Patrick. "Crowdsourcing the Evaluation of Post-Sandy Building Damage Using Aerial Imagery." *IRevolutions* (blog), November 1, 2012. https://irevolutions.org/2012/11/01/crowdsourcing-sandy-building-damage/.

Meier, Patrick. "Human Computation for Disaster Response." In *Handbook of Human Computation*, 95–104. New York: Springer, 2013. https://doi.org/10.1007/978-1-4614-8806-4_11.

Meier, Richard L. *A Communications Theory of Urban Growth*. Cambridge, MA: MIT Press, 1962.

Melgarejo-Heredia, Rafael, Leslie Carr, and Susan Halford. "The Public Web and the Public Good." In *Proceedings of the 8th ACM Conference on Web Science*, 330–332. WebSci '16. New York: ACM, 2016. https://doi.org/10.1145/2908131.2908181.

Meller, Helen. *Patrick Geddes: Social Evolutionist and City Planner*. London: Routledge, 2005.

Miami City Planning and Zoning Board. "The Miami Long Range Plan: Report on Tentative Plan for Trafficways." Miami, 1955.

"Microsoft CityNext: Coming to a City near You." Stories, July 10, 2013. https://news.microsoft.com/2013/07/10/microsoft-citynext-coming-to-a-city-near-you/.

M'ikanatha, Nkuchia M., Dale D. Rohn, Corwin Robertson, Christina G. Tan, John H. Holmes, Allen R. Kunselman, Catherine Polachek, and Ebbing Lautenbach. "Use of the Internet to Enhance Infectious Disease Surveillance and Outbreak Investigation." *Biosecurity and Bioterrorism: Biodefense Strategy, Practice, and Science* 4, no. 3 (September 1, 2006): 293–300. https://doi.org/10.1089/bsp.2006.4.293.

Miller, Greg. "1885 Map Reveals Vice in San Francisco's Chinatown and Racism at City Hall." *Wired*, September 30, 2013. https://www.wired.com/2013/09/1885-map-san-francisco-chinatow/.

Milne-Skinner, Andrew. "Liverpool's Slavery Museum: A Blessing or a Blight?'." *Racism, Slavery, and Literature (Frankfurt/M.: Peter Lang 2010)*, 2010, 183–195.

Mohl, Raymond A. "Stop the Road: Freeway Revolts in American Cities." *Journal of Urban History* 30, no. 5 (2004): 674–706.

Mohl, Raymond A. "Stop the Road: Freeway Revolts in American Cities." *Journal of Urban History* 30, no. 5 (July 2004): 674–706. https://doi.org/10.1177/0096144204265180.

Morrow, Nathan, Mock, Nancy, Papendieck, Adam, and Kocmich, Nicholas. "Independent Evaluation of the Ushahidi Haiti Project," 2011. https://www.researchgate.net/profile/Nathan_Morrow/publication/265059793_Ushahidi_Haiti_Project_Evaluation_Independent_Evaluation_of_the_Ushahidi_Haiti_Project/links/5451ef8f0cf2bf864cbaaca9/Ushahidi-Haiti-Project-Evaluation-Independent-Evaluation-of-the-Ushahidi-Haiti-Project.pdf.

Neis, Pascal, and Alexander Zipf. "Analyzing the Contributor Activity of a Volunteered Geographic Information Project—The Case of OpenStreetMap." *ISPRS International Journal of Geo-Information* 1, no. 2 (September 2012): 146–165. https://doi.org/10.3390/ijgi1020146.

"New Memo Reveals Census Question Was Added to Boost White Voting Power." Accessed June 26, 2019. https://slate.com/news-and-politics/2019/05/census-memo-supreme-court-conservatives-white-voters-alito.html.

New York City Department of Health and Mental Hygiene. "The New York City Community Air Survey: Neighborhood Air Quality 2008–2014," April 2016. https://www1.nyc.gov/assets/doh/downloads/pdf/environmental/comm-air-survey-08-14.pdf.

"*New York Times* Excludes Asian-Americans from Affirmative Action Study." *Daily Caller*. January 7, 2019. https://dailycaller.com/2014/04/23/new-york-times-excludes-asian-americans-from-affirmative-action-study/.

Norheim-Hagtun, Ida, and Patrick Meier. "Crowdsourcing for Crisis Mapping in Haiti." *Innovations: Technology, Governance, Globalization* 5, no. 4 (2010): 81–89.

NYCLU. "NYPD STOP-AND-FRISK ACTIVITY IN 2011 (2012)," 2012. https://www.nyclu.org/en/publications/report-nypd-stop-and-frisk-activity-2011-2012.

"NYCLU_2011_Stop-and-Frisk_Report.Pdf." Accessed January 10, 2019. https://www.nyclu.org/sites/default/files/publications/NYCLU_2011_Stop-and-Frisk_Report.pdf.

"NYPD Stop-and-Frisk Data." Accessed January 10, 2019. https://people.cs.uct.ac.za/~mkuttel/VISProjects2016/4AllieEtAl/index.html.

Obama, Barack. "Transparency and Open Government." Memorandum for the Heads of Executive Departments and Agencies, January 21, 2009. https://obamawhitehouse.archives.gov/the-press-office/transparency-and-open-government.

Oberholtzer, Jason. "Stop-And-Frisk By The Numbers." *Forbes*, 2012. https://www.forbes.com/sites/jasonoberholtzer/2012/07/17/stop-and-frisk-by-the-numbers/#5a02ed656703.

"Off the Map." *Economist*, November 13, 2014. https://www.economist.com/international/2014/11/13/off-the-map.

Offenhuber, Dietmar, and Carlo Ratti. *Waste Is Information: Infrastructure Legibility and Governance*. Cambridge, MA: MIT Press, 2017.

O'Kane, Sean. "Tesla and Waymo Are Taking Wildly Different Paths to Creating Self-Driving Cars." The Verge, April 19, 2018. https://www.theverge.com/transportation/2018/4/19/17204044/tesla-waymo-self-driving-car-data-simulation.

Olson, Donald R., Kevin J. Konty, Marc Paladini, Cecile Viboud, and Lone Simonsen. "Reassessing Google Flu Trends Data for Detection of Seasonal and Pandemic Influenza: A Comparative Epidemiological Study at Three Geographic Scales." *PLOS Computational Biology* 9, no. 10 (October 17, 2013): e1003256. https://doi.org/10.1371/journal.pcbi.1003256.

OpenStreetMap contributors. "Planet Dump Retrieved from Https://Planet.Osm.Org," 2017.

Organisation for Economic Cooperation and Development (OECD). "OECD Guidelines on the Protection of Privacy and Transborder Flows of Personal Data," September 23, 1980. http://www.oecd.org/internet/ieconomy/oecdguidelinesontheprotectionofprivacyandtransborderflowsofpersonaldata.htm.

Osborne, Thomas, and Nikolas Rose. "Spatial Phenomenotechnics: Making Space with Charles Booth and Patrick Geddes." *Environment and Planning D: Society and Space* 22, no. 2 (April 1, 2004): 209–228. https://doi.org/10.1068/d325t.

Palmer, J. J. N. "WARWICKSHIRE, PAGE 1." Open Domesday, n.d. https://opendomesday.org/book/warwickshire/01/.

Pappalardo, L., D. Pedreschi, Z. Smoreda, and F. Giannotti. "Using Big Data to Study the Link between Human Mobility and Socio-Economic Development." In *2015 IEEE International Conference on Big Data (Big Data)*, 871–878, 2015. https://doi.org/10.1109/BigData.2015.7363835.

Park, Robert E., and Ernest W. Burgess. *The City*. Chicago: University of Chicago Press, 2012.

Parrish, Melissa, G. Sarah, R. Emily, and W. Jennifer. "Location-Based Social Networks: A Hint of Mobile Engagement Emerges." *Forrester Research*, 2010.

Pastor-Escuredo, David, Alfredo Morales-Guzmán, Yolanda Torres-Fernández, Jean-Martin Bauer, Amit Wadhwa, Carlos Castro-Correa, Liudmyla Romanoff, et al. "Flooding through the Lens of Mobile Phone Activity." *IEEE Global Humanitarian Technology Conference (GHTC 2014)*. October 2014, 279–286. https://doi.org/10.1109/GHTC.2014.6970293.

Paulos, Eric, Richard J. Honicky, and Elizabeth Goodman. "Sensing Atmosphere." *Human-Computer Interaction Institute*, 2007, 203.

Pennington, Mark. "A Hayekian Liberal Critique of Collaborative Planning." *Planning Futures: New Directions for Planning Theory*, 2002, 187–208.

Peters, Adam. "This Device Creates A 3-D Soundscape To Help Blind People Navigate Through Cities." *Fast Company*, November 20, 2014. https://www.fastcompany.com/3038691/this-device-creates-a-3d-soundscape-to-help-blind-people-navigate-through-cities.

Peters, Jeremy W. "BP and Officials Block Some Coverage of Gulf Oil Spill." *New York Times*, June 9, 2010. https://www.nytimes.com/2010/06/10/us/10access.html.

Peterson, Jon A. *The Birth of City Planning in the United States, 1840–1917*. Baltimore: Johns Hopkins University Press, 2003.

Pickles, J. "Arguments, Debates, and Dialogues: The GIS–Social Theory Debate and the Concern for Alternatives." In Paul. A. Longley, Michael F. Goodchild, D. J. MacGuire, and David W. Rhind (eds.). *Geographical Information Systems: Principles, Techniques, Applications, and Management* (2nd ed.). New York: John Wiley and Sons, 2005, 12.

Pickles, John. *Ground Truth: The Social Implications of Geographic Information Systems*. Guilford Press, 1995.

Playfair, William. *Commercial and Political Atlas and Statistical Breviary (Original Version Was Published in 1786)*. Cambridge: Cambridge University Press, 2005.

"Public Lab: A DIY Environmental Science Community," 2019. https://publiclab.org/.

Public Lab Contributors. "Balloon Mapping Kit," n.d. https://store.publiclab.org/collections/featured-kits/products/mini-balloon-mapping-kit?variant=48075678607.

Purtova, Nadezhda. "Illusion of Personal Data as No One's Property." SSRN Scholarly Paper. Rochester, NY: Social Science Research Network, October 29, 2013. https://papers.ssrn.com/abstract=2346693.

Quercia, Daniele, Rossano Schifanella, Luca Maria Aiello, and Kate McLean. "Smelly Maps: The Digital Life of Urban Smellscapes." ArXiv:1505.06851, May 26, 2015. http://arxiv.org/abs/1505.06851.

Rambaldi, Giacomo. "Who Owns the Map Legend?" *URISA Journal* 17, no. 1 (2005): 5–13.

Ramirez, Jessica. "'Ushahidi' Technology Saves Lives in Haiti and Chile." Newsweek, March 3, 2010. https://www.newsweek.com/ushahidi-technology-saves-lives-haiti-and-chile-210262.

Reddy, Sasank, Deborah Estrin, and Mani Srivastava. "Recruitment Framework for Participatory Sensing Data Collections." In *International Conference on Pervasive Computing*, 138–155. New York: Springer, 2010.

Reinsel, David, John Gantz, and John Rydning. "The Digitization of the World From Edge to Core." Accessed January 25, 2019. https://www.seagate.com/files/www-content/our-story/trends/files/idc-seagate-dataage-whitepaper.pdf.

"Reporter Shows The Links Between The Men Behind Brexit And The Trump Campaign." NPR.org. Accessed July 6, 2019. https://www.npr.org/2018/07/19/630443485/reporter-shows-the-links-between-the-men-behind-brexit-and-the-trump-campaign.

Resch, Bernd, Anja Summa, Peter Zeile, and Michael Strube. "Citizen-Centric Urban Planning through Extracting Emotion Information from Twitter in an Interdisciplinary Space-Time-Linguistics Algorithm." *Urban Planning* 1, no. 2 (2016): 114–127. https://doi.org/10.17645/up.v1i2.617.

Rogstadius, J., M. Vukovic, C. A. Teixeira, V. Kostakos, E. Karapanos, and J. A. Laredo. "CrisisTracker: Crowdsourced Social Media Curation for Disaster Awareness." *IBM Journal of Research and Development* 57, no. 5 (September 2013): 4:1–4:13. https://doi.org/10.1147/JRD.2013.2260692.

Rosen, Sherwin. "Wage-Based Indexes of Urban, Quality of Life." *Current Issues in Urban Economics*, 1979, 74–104.

Rosenberg, Matthew. "Academic behind Cambridge Analytica Data Mining Sues Facebook for Defamation." *New York Times*, March 15, 2019. https://www.nytimes.com/2019/03/15/technology/aleksandr-kogan-facebook-cambridge-analytica.html.

Ross, Zev, Kazuhiko Ito, Sarah Johnson, Michelle Yee, Grant Pezeshki, Jane E Clougherty, David Savitz, and Thomas Matte. "Spatial and Temporal Estimation of Air Pollutants in New York City: Exposure Assignment for Use in a Birth Outcomes Study." *Environmental Health* 12 (June 27, 2013): 51. https://doi.org/10.1186/1476-069X-12-51.

Rothstein, Richard. "The Making of Ferguson." *Journal of Affordable Housing and Community Development Law* 24 (2015): 165–204.

Rudwick, Elliott M. "WEB Du Bois and the Atlanta University Studies on the Negro." In *WEB Du Bois*, 63–74. New York: Routledge, 2017.

Ruths, Derek, and Jürgen Pfeffer. "Social Media for Large Studies of Behavior." *Science* 346, no. 6213 (November 28, 2014): 1063–1064. https://doi.org/10.1126/science.346.6213.1063.

Ryzik, Melena. "'The Geography of Buzz,' a Study on the Urban Influence of Culture." *New York Times*, April 6, 2009. https://www.nytimes.com/2009/04/07/arts/design/07buzz.html.

Safecast. "Safecast Tile Map," 2019. http://safecast.org/tilemap/.

Safecast, and Marc Rollins. "Safecast Blog Image Compilation," n.d. https://blog.safecast.org/.

Sanborn Map Company. "Image 3 of Sanborn Fire Insurance Map from Boston, Suffolk County, Massachusetts." 5 (1888). http://hdl.loc.gov/loc.gmd/g3764bm.g03693188805.

Sanborn Map Company, and David Hodenfield. "Description and Utilization of the Sanborn Map (Compiled from the 1940 and 1953 Editions)." FIMo—How to Interpret Sanborn Maps, 2019. http://www.historicalinfo.com/fimo-interpret-sanborn-maps/#Utilization.

Sawicki, David S., and William J. Craig. "The Democratization of Data: Bridging the Gap for Community Groups." *Journal of the American Planning Association* 62, no. 4 (December 31, 1996): 512–523. https://doi.org/10.1080/01944369608975715.

Scott, James C. *Seeing like a State: How Certain Schemes to Improve the Human Condition Have Failed*. New Haven: Yale University Press, 1998.

"SDGs: Sustainable Development Knowledge Platform." Accessed July 14, 2019. https://sustainabledevelopment.un.org/sdgs.

Semuels, Alana. "The Internet Is Enabling a New Kind of Poorly Paid Hell." *Atlantic*, January 23, 2018. https://www.theatlantic.com/business/archive/2018/01/amazon-mechanical-turk/551192/.

Serajuddin, Umar, Hiroki Uematsu, Christina Wieser, Nobuo Yoshida, and Andrew Dabalen. *Data Deprivation: Another Deprivation to End*. Wahington, DC: World Bank, 2015.

Shah, Nayan. *Contagious Divides: Epidemics and Race in San Francisco's Chinatown*. Vol. 7. Berkeley: University of California Press, 2001.

Shaw, Clifford. "Map No. III Showing Addresses of 8591 Alleged Male Juvenile Delinquents Dealt with by the Juvenile Police Probation Officers during the Year 1927, Ten to Seventeen Years of Age / Prepared by Research Sociologists; Behavior Research Fund, Chicago." 1929. https://luna.lib.uchicago.edu/luna/servlet/view/search?q=_luna_media_exif_filename=G4104-C6E625-1927-S5.tif.

Shepard, Wade. *Ghost Cities of China: The Story of Cities without People in the World's Most Populated Country*. London: Zed Books, 2015.

Singer, Natasha. "I.B.M. Takes 'Smarter Cities' Concept to Rio de Janeiro." *New York Times*, March 3, 2012. https://www.nytimes.com/2012/03/04/business/ibm-takes-smarter-cities-concept-to-rio-de-janeiro.html.

Singer, Natasha, and Prashant S. Rao. "U.K. vs. U.S.: How Much of Your Personal Data Can You Get?" *New York Times*, May 20, 2018. https://www.nytimes.com/interactive/2018/05/20/technology/what-data-companies-have-on-you.html.

Snow, John. *On the Mode of Communication of Cholera*. London: John Churchill, 1855.

Sorace, Christian, and Hurst, William. "China's Phantom Urbanisation and the Pathology of Ghost Cities." *Journal of Contemporary Asia* 46, no. 2 (2016): 304–322.

Sotomayor, Sonia. *Schuette v. Coalition to Defend Affirmative Action*, US Supreme Court, No. 572 U.S. 2014.

Stanziola, Phil. *Mrs. Jane Jacobs, Chairman of the Comm. to Save the West Village Holds up Documentary Evidence at Press Conference at Lions Head Restaurant at Hudson & Charles Sts / World Telegram & Sun Photo by Phil Stanziola*. December 5, 1961. Photographic Print. LC-USZ-62–137838. Library of Congress. https://www.loc.gov/pictures/item/2008677538/.

Starbird, Kate, and Leysia Palen. *Pass It On?: Retweeting in Mass Emergency*. International Community on Information Systems for Crisis Response and Management, 2010.

Starbird, Kate, Leysia Palen, Amanda L. Hughes, and Sarah Vieweg. "Chatter on the Red: What Hazards Threat Reveals about the Social Life of Microblogged Information." In *Proceedings of the 2010 ACM Conference on Computer Supported Cooperative Work*, 241–250. ACM, 2010.

Steffel, R. Vladimir. "The Boundary Street Estate: An Example of Urban Redevelopment by the London County Council, 1889–1914." *Town Planning Review* 47, no. 2 (1976): 161–173.

Stempeck, Matt. "Sharing Data Is a Form of Corporate Philanthropy." *Harvard Business Review*, July 24, 2014. https://hbr.org/2014/07/sharing-data-is-a-form-of-corporate-philanthropy.

Stieglitz, C. M. *Sponsor of Battery Bridge / World Telegram & Sun Photo by C. M. Stieglitz*. 1939. Photographic print. NYWTS—BIOG—Moses, Robert—Ex-Parks Commissioner [item] [P&P]. Library of Congress Prints and Photographs Division Washington, DC.

Stiftelsen Flowminder. "Flowminder.Org World Pop Map," 2015. https://web.flowminder.org/worldpop.

Stiglitz, Joseph E. *Knowledge as a Global Public Good*. Oxford: Oxford University Press, 1999. https://www.oxfordscholarship.com/view/10.1093/0195130529.001.0001/acprof-9780195130522-chapter-15.

"'Stowage of the British Slave Ship "Brookes" under the Regulated Slave Trade, Act of 1788.'" The Abolition Seminar. Accessed July 19, 2019. https://www.abolitionseminar.org/brooks/.

Sui, Daniel, Sarah Elwood, and Michael Goodchild. *Crowdsourcing Geographic Knowledge: Volunteered Geographic Information (VGI) in Theory and Practice*. New York: Springer Science & Business Media, 2012.

Sultana, Selima. "Job/Housing Imbalance and Commuting Time in the Atlanta Metropolitan Area: Exploration of Causes of Longer Commuting Time." *Urban Geography* 23, no. 8 (December 1, 2002): 728–749. https://doi.org/10.2747/0272-3638.23.8.728.

Swabey, Pete. "IBM, Cisco and the Business of Smart Cities." *Information Age*, February 23, 2012. https://www.information-age.com/ibm-cisco-and-the-business-of-smart-cities-2087993/.

Swaine, Jon. "Police Will Be Required to Report Officer-Involved Deaths under New US System." *Guardian*, August 8, 2016. https://www.theguardian.com/us-news/2016/aug/08/police-officer-related-deaths-department-of-justice.

Taddeo, Mariarosaria. "Data Philanthropy and the Design of the Infraethics for Information Societies." *Philosophical Transactions of the Royal Society A* 374, no. 2083 (December 28, 2016): 20160113. https://doi.org/10.1098/rsta.2016.0113.

Tang, Mingzhe, and N. Edward Coulson. "The Impact of China's Housing Provident Fund on Homeownership, Housing Consumption and Housing Investment." *Regional Science and Urban Economics* 63 (2017): 25–37.

Taylor, Linnet. "The Ethics of Big Data as a Public Good: Which Public? Whose Good?" SSRN Scholarly Paper. Rochester, NY: Social Science Research Network, August 9, 2016. https://papers.ssrn.com/abstract=2820580.

Taylor, Linnet. "The Ethics of Big Data as a Public Good: Which Public? Whose Good?" *Philosophical Transactions of the Royal Society A* 374, no. 2083 (2016): 20160126.

Texas Department of Transportation (TxDOT). "Texas US House Districts." TxDOT Open Data Portal, 2019. https://gis-txdot.opendata.arcgis.com/datasets/texas-us-house-districts?geometry=-334.512%2C-52.268%2C334.512%2C52.268.

Thatcher, Jim, David O'Sullivan, and Dillon Mahmoudi. "Data Colonialism through Accumulation by Dispossession: New Metaphors for Daily Data." *Environment and Planning D: Society and Space* 34, no. 6 (December 1, 2016): 990–1006. https://doi.org/10.1177/0263775816633195.

"The Brookes—Visualising the Transatlantic Slave Trade." Accessed July 19, 2019. https://www.history.ac.uk/1807commemorated/exhibitions/museums/brookes.html.

"The Counted: Tracking People Killed by Police in the United States." *Guardian*, 2016. https://www.theguardian.com/us-news/series/counted-us-police-killings.

"The Data Deluge." *Economist*, February 25, 2010. https://www.economist.com/leaders/2010/02/25/the-data-deluge.

"The Next Chapter for Flu Trends." *Google AI Blog* (blog), October 15, 2018. http://ai.googleblog.com/2015/08/the-next-chapter-for-flu-trends.html.

Thrasher, Frederic M. *The Gang: A Study of 1,313 Gangs in Chicago.* Chicago: University of Chicago Press, 1927.

Tiecke, Tobias. "Open Population Datasets and Open Challenges." Facebook Code. n.d. https://code.fb.com/core-data/open-population-datasets-and-open-challenges/.

Topalov, Christian. "The City as Terra Incognita: Charles Booth's Poverty Survey and the People of London, 1886–1891." *Planning Perspectives* 8, no. 4 (October 1, 1993): 395–425. https://doi.org/10.1080/02665439308725782.

Townsend, Anthony M. *Smart Cities: Big Data, Civic Hackers, and the Quest for a New Utopia.* New York: Norton, 2013.

University of Chicago. "Social Base Map of Chicago: Showing Industrial Areas, Parks, Transportation, and Language Groups / Prepared by the University of Chicago Local Community Research Committee." 1926. http://pi.lib.uchicago.edu/1001/cat/bib/39315.

"Ushahidi." Accessed January 25, 2019. https://www.ushahidi.com/.

Vallianatos, Mark. "Uncovering the Early History of 'Big Data' and the 'Smart City' in Los Angeles." *Boom California* (blog), June 16, 2015. https://boomcalifornia.com/2015/06/16/uncovering-the-early-history-of-big-data-and-the-smart-city-in-la/.

Varian, Hal R. *Markets for Information Goods*. Vol. 99. Citeseer, 1999.

Varian, Hal R. *Microeconomic Analysis, Third Edition*. 3rd edition. New York: Norton, 1992.

Vaughan, Laura. "The Charles Booth Maps as a Mirror to Society." *Mapping Urban Form and Society* (blog), October 3, 2017. https://urbanformation.wordpress.com/2017/10/03/the-charles-booth-maps-as-a-mirror-to-society/.

Vicens, Natasha. "How One Resident near Fracking Got the EPA to Pay Attention to Her Air Quality." PublicSource, December 15, 2016. https://www.publicsource.org/how-one-resident-near-fracking-got-the-epa-to-pay-attention-to-her-air-quality/.

Vieweg, Sarah, Amanda L. Hughes, Kate Starbird, and Leysia Palen. "Microblogging during Two Natural Hazards Events: What Twitter May Contribute to Situational Awareness." In *Proceedings of the SIGCHI Conference on Human Factors in Computing Systems*, 1079–1088. ACM, 2010.

Waddell, Paul, and Gudmundur F. Ulfarsson. *Accessibility and Agglomeration: Discrete-Choice Models of Employment Location by Industry Sector*. Vol. 63. 82nd Annual Meeting of the Transportation Research Board, Washington, DC, 2003.

Wang, Fahui, and W. William Minor. "Where the Jobs Are: Employment Access and Crime Patterns in Cleveland." *Annals of the Association of American Geographers* 92, no. 3 (2002): 435–450. https://doi.org/10.1111/1467-8306.00298.

Watts, Jonathan. "Almost Four Environmental Defenders a Week Killed in 2017." *Guardian*, February 2, 2018. https://www.theguardian.com/environment/2018/feb/02/almost-four-environmental-defenders-a-week-killed-in-2017.

Watts, Jonathan. "China Prays for Olympic Wind as Car Bans Fail to Shift Beijing Smog." *Guardian* 21 (2007).

Weizman, Eyal. "Introduction: Forensis." *Forensis: The Architecture of Public Truth*, 2014, 9–32.

Wekesa, Gilbert Koech and Chrispinus. "Kenya: Nairobi Transit Map Launched." *Star*, January 29, 2014. https://allafrica.com/stories/201401290479.html.

Welter, Volker M, and Whyte, Iain Boyd. *Biopolis: Patrick Geddes and the City of Life*. Cambridge, MA: MIT Press, 2002.

Wesolowski, Amy, Caroline O. Buckee, Linus Bengtsson, Erik Wetter, Xin Lu, and Andrew J. Tatem. "Commentary: Containing the Ebola Outbreak—the Potential and Challenge of Mobile Network Data." *PLoS Currents* 6 (September 29, 2014). https://doi.org/10.1371/currents.outbreaks.0177e7fcf52217b8b634376e2f3efc5e.

Wesolowski, Amy, Nathan Eagle, Andrew J. Tatem, David L. Smith, Abdisalan M. Noor, Robert W. Snow, and Caroline O. Buckee. "Quantifying the Impact of Human Mobility on Malaria." *Science* 338, no. 6104 (October 12, 2012): 267–270. https://doi.org/10.1126/science.1223467.

"What Is Responsible Data?" *Responsible Data* (blog). Accessed July 28, 2019. https://responsibledata.io/what-is-responsible-data/.

"What We Can Learn From the Epic Failure of Google Flu Trends," *Wired*, October 17, 2018. https://www.wired.com/2015/10/can-learn-epic-failure-google-flu-trends/.

Wilbur Smith and Associates. "A Major Highway Plan for Metropolitan Dade County, Florida." New Haven, CT: Prepared for State Road Department of Florida and Dade County Commission, 1956.

Wilbur Smith and Associates. "Alternates for Expressways: Downtown Miami, Dade County, Florida." New Haven, CT, 1962.

Williams, Sarah. "Here Now! Social Media and the Psychological City." In *Inscribing a Square— Urban Data as Public Space*. New York: Springer, 2012.

Williams, Sarah, Erica Deahl, Laurie Rubel, and Vivian Lim. "City Digits: Local Lotto: Developing Youth Data Literacy by Investigating the Lottery." *Journal of Digital and Media Literacy*, 2015.

Williams, Sarah, Jacqueline Klopp, Peter Waiganjo, Daniel Orwa, and Adam White. "Digital Matatus: Using Mobile Technology to Visualize Informality." *Proceedings ACSA 103rd Annual Meeting: The Expanding Periphery and the Migrating Center*, 2015.

Williams, Sarah, Elizabeth Marcello, and Jacqueline M. Klopp. "Toward Open Source Kenya: Creating and Sharing a GIS Database of Nairobi." *Annals of the Association of American Geographers* 104, no. 1 (January 2, 2014): 114–130. https://doi.org/10.1080/00045608.2013.846157.

Williams, Sarah, Wenfei Xu, Shin Bin Tan, Michael J. Foster, and Changping Chen. "Ghost Cities of China: Identifying Urban Vacancy through Social Media Data." *Cities* 94 (2019): 275–285.

Williamson, Jeffrey G. *Coping with City Growth during the British Industrial Revolution*. Cambridge: Cambridge University Press, 2002.

Wilson, Robin, Elisabeth zu Erbach-Schoenberg, Maximilian Albert, Daniel Power, Simon Tudge, Miguel Gonzalez, Sam Guthrie, et al. "Rapid and Near Real-Time Assessments of Population Displacement Using Mobile Phone Data Following Disasters: The 2015 Nepal Earthquake." *PLoS Currents* 8 (February 24, 2016). https://doi.org/10.1371/currents.dis.d073f bece328e4c39087bc086d694b5c.

Wines, Michael. "Deceased G.O.P. Strategist's Hard Drives Reveal New Details on the Census Citizenship Question." *New York Times*, May 30, 2019. https://www.nytimes.com/2019/05/30/ us/census-citizenship-question-hofeller.html.

World Bank. "Open Traffic Data to Revolutionize Transport," December 19, 2019. http://www .worldbank.org/en/news/feature/2016/12/19/open-traffic-data-to-revolutionize-transport.

World Bank. "World Bank Group and GSMA Announce Partnership to Leverage IoT Big Data for Development," February 26, 2018. http://www.worldbank.org/en/news/press-release/2018/02/26/world-bank-group-and-gsma-announce-partnership-to-leverage -iot-big-data-for-development.

Wu, Jing, Joseph Gyourko, and Yongheng Deng. "Evaluating the Risk of Chinese Housing Markets: What We Know and What We Need to Know." *China Economic Review* 39 (July 1, 2016): 91–114. https://doi.org/10.1016/j.chieco.2016.03.008.

Yao, Y., and Y. Li. "House Vacancy at Urban Areas in China with Nocturnal Light Data of DMSP-OLS." In *Proceedings 2011 IEEE International Conference on Spatial Data Mining and Geographical Knowledge Services*, 457–462, 2011. https://doi.org/10.1109/ICSDM.2011.5969087.

Yardley, Jim. "Beijing's Olympic Quest: Turn Smoggy Sky Blue." *New York Times*, December 29, 2007.

Zheng, Siqi, Yuming Fu, and Hongyu Liu. "Demand for Urban Quality of Living in China: Evolution in Compensating Land-Rent and Wage-Rate Differentials." *Journal of Real Estate Finance and Economics* 38, no. 3 (April 1, 2009): 194–213. https://doi.org/10.1007/s11146-008-9152-0.

Zheng, Siqi, and Matthew E. Kahn. "Land and Residential Property Markets in a Booming Economy: New Evidence from Beijing." *Journal of Urban Economics* 63, no. 2 (March 1, 2008): 743–57. https://doi.org/10.1016/j.jue.2007.04.010.

Zhou, Pengfei, Yuanqing Zheng, and Mo Li. "How Long to Wait?: Predicting Bus Arrival Time with Mobile Phone Based Participatory Sensing." In *Proceedings of the 10th International Conference on Mobile Systems, Applications, and Services*, 379–392. ACM, 2012.

Zook, Matthew, Solon Barocas, danah boyd, Kate Crawford, Emily Keller, Seeta Peña Gangadharan, Alyssa Goodman, et al. "Ten Simple Rules for Responsible Big Data Research." *PLOS Computational Biology* 13, no. 3 (March 30, 2017): e1005399. https://doi.org/10.1371/journal.pcbi.1005399.

INDEX

Note: Page numbers in italics indicate figures; page numbers in bold indicate tables.

Goodchild, Michael, 54
Google, 94, 130–131, 188, 193
  augmented reality navigation tool, 190, *190*
  data collection and, 189
  Google Assistant, 94
  Google Earth, 71
  Google Flu Trends (GFT), 130–131, 134, 215
  Google Maps, 54–55, 69, 95, 146
  Google Maps transit directions, 77, 149
  Google Purchases, 94
  Sidewalk Labs, 47
  Waymo (autonomous vehicle program), 189
Govern, Maureen, 197
Government(s). *See also specific countries*
  collection of data from citizens, 78–79
  protection from surveillance by, xix
  release of data and, 69
GPS data for location, 52, 80, 204
Grab Taxi, 204, *205*, 206
Gray Area, 85
Great Britain, 14, 91. *See also* Brexit; London, England
  Ordnance Survey, 10, 14, 54
Great Depression, 36
"Great Society," 43
Greenhost, 199
Greenwich Village, New York City, 39
Ground truthing, xx, 218–219
Guardian, 56, 175–176, 184
Gulf of Mexico, British Petroleum (BP) oil spill, 69, 70

Hackathon, University of Nairobi, 152, *152*
Hackers: Heroes of the Computer Revolution (Levy), 89
Hacking, 89–136. *See also* Hackathon, University of Nairobi
  creative, xv–xvi
  ethical and responsible "white hacks," 133–135
  hacker culture, 89–90
Haiti, 200, 203
  2010 earthquake in, 81, 129, 132–133
Hall, Peter, 43
Hangzhou, China, 98

Hansen's gravitational model of spatial accessibility, 101, *102*
Harley, Brian, 155–156
Harvard University, 176
Harvey, David, 156
Haselton, Todd, 94
Healthmap.org, 129–130
"Here Now! Social Media and the Psychological City" project, 121, *123*
Hillier, Amy E., 36
Hispanic students, percentage accepted to state universities, 174
Hofeller, Thomas B., 6–7
Home Owners' Loan Corporation (HOLC), 34, *35*, 42, 220
Hotlines, data collection through, 128–129, 135
Housing Act of 1949, 36
Housing Provident Fund (China), 97
Howe, Jeff, 73
Hull House, Chicago, 22
Hull House Maps and Papers (Kelley and Adams), 22, *23*
Humanitarian OpenStreetMap, 80, 82, *83–84*
Humanitarian Tracker, 130
Huridocs, 199
Hurricane Sandy, mapping damage after, 82–84

IBM, 5, 47–48
Image of the City, The (Lynch), 120
Incan fiber recording device (khipu), 2, *4*
Incarceration, Million Dollar Blocks project and, 158, *159*, *160*
Industrial Revolution, xiv, 9, 10, 129
Instagram, 139
Institute for Transportation and Development Policy (ITDP), 153
International Charter on Space and Major Disasters (1999), 209
International Data Corporation (IDC), xvi, 51–52
International Olympics Committee, 57
International Research Board (IRB), 92
Interstate 95 highway plan (Miami, Florida), *38*, 39

Urban modeling, rebooted, 47–48
Urban planners, 31–34, 36–37, 39–41, 155–156. *See also* Urban planning
Urban planning, 9–10, 31–32, 155–156
  in China, 108, 110
  data analytics and, 47–48
  population data and, 203–204
  racist reasoning and, 36–37, 39
  survey and mapping techniques, 42–43
Urban populations, 10
Urban renewal
  data and, 36–37, 39
  funds for, 34, 126
Urban Systems Laboratory, MIT (Massachusetts Institute of Technology), 44
USAID (United States Agency for International Development), 55
US Bill of Rights, 137
US Census, 5
  of 2010, 168
  African Americans and, 5
  citizenship question and, 5–6
US Census Bureau, 5
US Constitution, 5, 137, 180
US Department of Defense, 193
US Department of Transportation, 47
US Federal Trade Commission, "Fair Information Practices," 198
Ushahidi platform, 80–83, 199
US National Science Foundation, 193
US Supreme Court, 5, 174, 195

Value, extracting, 194
Varian, Hal R., 194
Victorian era, maps and, 14
Vision 2030 ICT Innovation Award, 152
Volunteered Geographic Information (VGI), 52, 54
Voting Rights Act of 1965, 5

Wall Street Journal, 178
War, technology of, 43–45
Warfare to Welfare (Light), 43
Warren, Jeff, 70
Washington, 174
Washington Post, 56–57

Waymo (Google's autonomous vehicle program), 189, 195
Waze, 206
Weapons of Math Destruction (O'Neil), xii
Websites, interactive, 164–174
Weibo, 98, 112–113
West Indies, 140
White, Jeremy, 175
Wichita, Kansas, Million Dollar Blocks project and, 162
Wiener, Norbert, 44
Wikipedia, 54
Wilbur Smith and Associates, 39
Williams, Sarah, x, 95, 96, 115, 126, 140, 187, 218
William the Conqueror (King William I), 2
WNYC, 179
Wood, Denis, 155
Works Progress Administration, Real Property Surveys, 36
World Bank, 78, 196, 204, 206–208
World Cities Summit (WCS) Mayors Forum 2019, 187
World Health Organization, 61, 62, 86, 198
World Resources Institute, 196
World That Counts, A (United Nations report), 188
Wuhan, China, 98

Yelp, 95

Zoning, 30–34, 36
Zoning commission, St. Louis, Missouri, 34